INTERNATIONAL ACADEMY OF SANSKRIT RESEARCH

VYMAANIKA-SHAASTRA
AERONAUTICS

by

MAHARSHI BHARADWAAJA

❖❖❖ ❖❖

Propounded by
Venerable SUBBARAYA SASTRY

Translated into English and
Edited, Printed and Published by
G. R. J O S Y E R
SCHOLAR, HISTORIAN, ESSAYIST, SANSKRITIST

Maharshi Bharadwaaja
Vymaanika-Shaastra or Science of Aeronautics
Edizioni L'Arcipelago – 178 pp.
ISBN 978-88-89517-18-5

"La redazione del testo è di Edizioni L'Arcipelago"

www.edizionilarcipelago.it

ISBN 978-88-89517-18-5

CONTENTS

FOREWORD

On 25-8-1952 the Mysore representative of the Press Trust of India, Sri N. N. Sastry, sent up the following report which was published in all the leading dailies of India, and was taken up by Reuter and other World Press News Services:

"Mr. G. R. Josyer, Director of the International Academy of Sanskrit Research in Mysore, in the course of an interview recently, showed some very ancient manuscripts which the Academy had collected. He claimed that the manuscripts were several thousands of years old, compiled by ancient rishis, Bharadwaja, Narada and others, dealing, not with the mysticism of ancient Hindu philosophy of Atman or Brahman, but with more mundane things vital for the existence of man and progress of nations both in times of peace and war.

"Mr. Josyer's manuscripts dealt in elaborate detail about food processing from various indigenous materials like grass, vegetables and leaves for human consumption, particularly during times of famine.

"One manuscript dealt with Aeronautics, construction of various types of aircraft for civil aviation and for warfare. He showed me plans prepared according to directions contained in the manuscript on Aeronautics of three types of aircraft or Vimanas, namely, Rukma, Sundara and Shakuna Vimanas. Five hundred slokas or stanzas dealing with these go into such intricate details about choice and preparation of metals that would be suitable for various parts of vimanas of different types, constructional details, dimensions, designs and weight they could carry, and purposes they could be used for.

"Mr. Josyer showed some types of designs and drawing of a helicopter-type cargo-loading plane, specially meant for carrying combustibles and ammunition, passenger aircraft carrying 400 to 500 persons, double and treble-decked aircraft. Each of these types had been fully described.

"In the section giving about preparation and choice of metals and other materials that should go into such construction of aircraft, details were specified that the aircraft, (these metals are of 16 different alloys), must be "unbreakable, which cannot be cut through, which would not catch fire, and cannotbe destroyed by accidents." Details as to how to make these vimanas in flight invisible through smoke screens are given in Vimanasastra of Maharshi Bharadwaja.

"Further description and method of manufacturing aircraft, which will enable pilots not only to spot enemy aircraft, but also to hear what enemy pilots in their planes were speaking, on principles akin to radar, have all been given in elaborate detail with suitable explanatory notes. There are eight chapters in this book which deal with construction of aircraft, which fly in air, go under water, or float on water.

TRAINING OF PILOTS

"A few slokas deal with qualifications and training of pilots to man these aircraft. These ancient types of aircraft are provided with necessary cameras to take pictures of approaching enemy planes. Yet another set of slokas deals with the kind of food and clothing to be provided for pilots to keep them efficient and fit in air flying conditions.

"Mr. Josyer said he was attempting to publish these manuscripts suitably translated in English.

"Another manuscript dealt with ancient Indian architecture, fully illustrated to facilitate construction. This treatise is ascribed to Maharshi Narada, and gives elaborate details about choice of constructional material for various types of buildings, even 15 storeys high. Sectional drawing has also been provided. A few chapters deal with construction of villages, cities and towns, fortresses, palaces and temples. This manuscript is full of plans and engineering constructional details to guide engineers.

"Yet another manuscript from which Mr. G. R. Josyer read out passages referred to preparation of imitation diamonds and pearls. He also showed me another remarkable manuscript which deals in detail about food processing for invalids, for youth and for old and debilitated persons."

A mild avalanche of letters blew towards us during the following days from all over India. One of the first was from James Burke of "Life International", from Delhi asking if he could come and see the Mss. We replied, "Please wire 1000 dollars, and then come." He was taken aback, and wrote that he thought people here felt honoured by being mentioned in "Life International," but that we seemed to be different. We did not reply. Now James Burke is dead; and great "Life International" also is no more! Such is human evanescence!

Miss Jean Lyon, journalist of Toronto and New York, wrote from Delhi that she would visit us. She came and saw the Mss, and recorded her interview with us in her book "Just Half a world Away" in a chapter headed "Science by Sutras", concluding with the charge that we were guilty of a rabid nationalism, seeking to wipe out everything since the Vedas!

That is no way for a journalist to judge persons. We only hold that for Indians, or others, to wipe out the Vedas is absurd! We are neither rabid, nor national. God has created the Earth like a round ball, all its contents forming one compact unit, not a hundred and odd broken units as in the League of Nations. Only in maps is the earth shown broken into 2 hemispheres. If you actually break the Earth into 2 hemi-spheres, you will be having Doomsday!

Ours is not nationalism, rabid or tame, but one world humanism, or world-citizenship. That should not shock Miss Jean Lyon.

Others who wrote to enquire with excited wonder were Governor R. R. Diwakar of Behar, Maharaja of Kashi, Dr. Trivedi of "Searchlight," Patna, Professor Dwivedi of Gorakhpur,

Professor Chauhan of Seoni, Professor Theeanee of Madras, Swamy Chaitanya of Mussoorie, H. R. Sharma of Phagwara, Harit Krishna Deb of Calcutta, R. B. Lal of Allahabad, P. S. Bharathi of Ootacamand, Miss McIntyre of Bangalore, M. V. Sharma, Industrialist, Madras, D. V. Potdar of Poona, Raja A. K. N. Singh of Ramnagar, U. P., Rao Bahadur M.A. Rangaswamy, Patna, N. Anandalwar, Bangalore.

S. M. Sharma, editor, wrote in "Searchlight," Patna, "To an eminent Sanskrit scholar, Shri G. R. Josyer, Director of the International Academy of Sanskrit Research, Mysore, we owe the discovery of the manuscript on Vimanas by Maharshi Bharadwaja. Shri Josyer is already more than 70. Many Will share my hope and prayer that the Nehru Goverment would lose no time in acquiring the manuscript, which, according to my information, is most unique." Editor Sharma too is no more!

The Mss. came to us by Divine grace! When on 28-6-1951 we got H. H. the Maharaja of Mysore to inaugurate the International Academy of Sanskrit Research, evidently it was an auspicious occasion. The Academy has truly attained International fame, and has become known from one end of the globe to the other!

One of the guests coming from Bangalore for the ceremony brought a small manuscript in exercise book form containing the beginning of Maharshi Bharadwaja's "Vymanika Sastra." We were struck by it, and exhibited it along with our other Mss. in various stages of decay, to H. H. The Maharaja and Chief Minister K. C. Reddy and others when we took them round.

After the function the Mss. was returned to the guest, who gave it back to the custodian of Pandit Subbaraya Sastry's literary records, Sri Venkatrama Sastry, B.A., B.L., Advocate of the Bangalore Bar. Subsequently we contacted him, and on our promise of doing our best to publish them, he was good enough to let us have copies of some of the manuscripts. The message of the Press Trust of India was with reference to them. The fan mail resulting therefrom brought enquiries from personages such as

Air Commodore Goyal of the Western Command, Bangalore, The Editor of the Kesari and Mahratta, Poona, Major Gadre of Saraswati Mahal, Tanjore, Minister A. G. Ramachandra Rao, Bangalore, Sri Swamy of Bhandarkeri Mutt, M. G. Seth, Bombay, P. D. Padam Chand, Delhi, P. M. Kabali, Bombay, Aeronautical Society of India, Ministry of Scientific and Cultural Affairs, Delhi, the Director General of Civil Aviation, the Hindustan Aircrafts, Ltd.

We then commenced printing the original in Sanskrit, and had made some progress, when suddenly there came a harsh letter from the donor, Sri Venkatrama Sastry, accusing us of exploiting the manuscripts for our personal benefit. Having had no such idea ourselves, it evoked disgust, and we replied that he could take back the manuscripts, and discontinued the printing!

But then letters continued to come from far away, from estimable men avidly interested in the manuscript, and organs of learned Associations and books by scholars of the abstruse began to quote us as intending to bring out the publication. Seigfried Hansch, Deschenes, Canada, Hans Krefft, Berlin, Blaes-Gustaf-Nordquist, Stockholm, Sweden, Bjorn Loven, Innsbruck, Austria, Joachim Rothaner, Kellerburg, Austria, Jan Wallgren, Stockholm, P. Salzmann de la Mar, Eskilstuna, Sweden, Hans-Werner-Von Engel, Bad Gadesberg, West Germany, Sten Lindgren, Stockholm, Lars Eric Helin, Kalender, Gothenburg, Sweden, WM. Dawson & Sons, London, Charles Danois, Kristianstad, Sweden, James Alves, Sao Paulo, Brazil, Torbjorn Holmquist, Vetlanda, Sweden, Ernest Heinrich, Homburg, Klaus Aarsleff Jorgensen, Skellingsted, Denmark, Gosta Karlsson, Stockholm, Peter Bernin, Malmo, Sweden, Dr. Curtis J Mccall, Lake Worth, Florida, Robert Ashley Falk, Auckland, New Zealand, Terry W. Colvin, Evansville, U.S.A., Sven Bertil Hansson, Malmo, Sweden, Kjell Ericson, Borlonge, Maurice T. Caison, North Carolina, M. A. Gresham, Jr. Atlanta, U.S.A. Alan Y. Wilcox, Lauderdale, Florida, Strubes, Copenhagen, Demmark, Alan D. William, Downey, California, Stuve Sundquist, Uppsala, Bo H. Svensson, Sweden, Karen Kesti, Republic WA, U.S.A., Richard Watson, New South Wales, Australia, Ontario

College of Art, Toronto, Canada, Dr. Jacq Eskens, Rotterdam, Netherlands, Bernin Co Moberg, Montmartre, Paris, L.S.U. Rydberg, Stockholm, Chairman, E. A. G. Mackay, British Unidentified Flying Object Research Association, London, Mrs. Annica Foxcraft, Transvaal, Oliver Williams, Weimar, Texas, Jan Swagermann, Ship "Johannra", Amsterdam, Robert B. Young, Jr. Radco Incorporated, Houston, Texas, Sam J. Lundwall, Askild and Karnekull, Stockholm, Dr. Cedric Leonard, Oklahoma, Gwendelholm, Stockholm, Michele Bonamici, Milan, Italy, Jose M. Fernandez, Goteborg, Sweden, President Lennort Lidfoss, Spectrum, Forlags, A. B. Danderyd, Sweden.

The "Clima Astral" of Brazil, "The Mexican News" of Mexico, The "Spectrum" and "Pursuit" of U.S.A. and three Swedish books on ancient and astral research, "Kulturer Fore Istiden" by Ivan Troing, "Forntidens Teknik" by H. Kjellson, and "Flygande Tefat" by Max B. Miller, reproduced our original descriptive leaflet, and announced that we expected to publish the remarkable volume. The Maharaja of Mysore invited us, and after offering Tea, wanted the Manuscript for being shown to Dr. Thacker, the National Scientist of India. We reminded him that Sanskritists were averse to parading knowledge before idle curiosity, and that the manuscript had to be translated into English and tested by research, and then only made available for the public gaze. Four scientist Doctors from the Indian Institute of Science, Bangalore, came with a letter of Introduction from the Head of The Department of Power Engineering, Dr. M.A. Tirunarayanan, and sought clarification from us!

It was as if the orange-shaped Earth had become a porcupine, and was shooting its quills at us from all sides, in order to goad us into the task, which we had been reluctant to take up as being too onerous for us!

We had therefore to gather strength from the cumulative good-will of the world-wide public, and gird ourselves up for the strenuous undertaking.

To be really of value the volume had to include the Sanskrit original, its translation in English as demanded by western readers, and sketches showing the designs of the Vimanas for further clarification. It had to be a beautiful volume commensurate with the magnificence of its subject, and the high expectations of the public awaiting its publication.

Thus, at the age of 81 we had to sit up and translate the technical Sanskrit into readable English, and scrutinise the printing of both the Sanskrit and English, involving the strain of multiple proof-reading. The finance required was considerable, and as no help was forthcoming, we had to scrape together the meagre savings of a life-time, procure needful printing equipment at mounting costs, engage labour at emergency rates, and at long last, with the help of Divine grace, are able to herald the birth of the volume, which has been in gestation for over ninety years!

"Vymaanika Shastra" consists of nearly 6000 lines, or 3000 verses of lucid Sanskrit, dealing with the construction of Vimaanas or Aeroplanes. That the vocabulary of ancient Sanskrit could in simple flowing verse depict the technical details with effortless ease is a tribute to the language, and the greatness of the author.

Maharshi Bharadwaja is an august name in the pantheon of Hindu Sages who recorded Indian civilization, in the spiritual, intellectual, and scientific fields in the hoary past. They transmitted knowledge from mouth to mouth, and from ear to ear, for long eras. Written transmission through birch-backs or palm-leaves, or home-made paper, are from this side of a thousand years. Even they are to be found in mangled forms owing to the depredation of time, weather and insect hordes. There is no-written material for the vast volume of Vedas, Upanishads, Shastras, and Puranas, which have come down for over 10000 years as a patrimony, not only for India, but for mankind in general. They remain imbedded in the ether of the sky, to be revealed—like television,—to gifted mediums of occult perception.

Venerable Pandit Subbaraya Sastry, who has left the legacy of manuscript treasures including "Vymanika Shastra", was a simple, orthodox, intellectual Brahmin with spiritual gifts, who was esteemed by all who knew him, Englishmen and anglicised or educated Indians, in various walks of life.

He was a walking lexicon gifted with occult perception. His sole aim was to transmit his knowledge to posterity. He lived a life of poverty, like Socrates, and sought no gains for himself.

In 1885 Mr. B. Suryanarain Rao, B.Sc., M. R. A. S., distinguished Astrologer and Editor, first met him and became his devoted exponent. In 1911 he started a Magazine in Madras named "Bhowthika Kalaa Nidhi," or "Treasure house of physical sciences", and published extracts from the revelations of the venerable scholar. We are in possession of 6 issues of that rare Journal which came to us by Divine grace.

On 1-8-1918 he began to dictate "Vymanika Sastra" to Mr. Venkatachala Sarma, who took down the whole in 23 exercise books up to 23-8-1923.

That gave manuscript shape to Maharshi Bharadwaja's "Vymanika Sastra". Then by a flash of genius he engaged a draughtsman, and got drawings of some varieties of the Vimanas prepared under his instructions, which form an indispensable adjunct to the manuscript proper. That was in 1923. India was then under British rule. Gandhi's Non-co-operation movement was catching fire. And, it is said, Pandit Subbaraya Sastry was arrested! Yeoman efforts procured his release. But his activities had to remain confined. In 1928 he addressed a letter to the Maharaja of Darbbanga for aid in publication of the manuscripts. But the rich in India have got deaf ears, and warped minds! Then, disappointed and broken-hearted, in the early 30's, venerable Subbaraya Sastry passed out of this world, and left it the poorer thereby!

For some 20 years his literary treasures remained as under frigidaire, guarded by his daughter and young Mr. Venkatrama Sastry. Then the Unseen Powers began to play, and the manuscripts were released to light. And at last it has pleased God to enable us to present Maharshi Bharadwaja's "Vymanika Sastra" to the world's elite, and pay our tribute to the memories of Maharshi Bharadwaja and venerable Subbaraya Sastry.

We thank God for His gracious favour. We thank Mr. Venkatrama Sastry who made the manuscript available to us; our first son, G. S. Josyer, M.A., B.ED., who contacted Mr. Venkatrama Sastry and brought the Mss., prepared it for the press, and even composed a portion of the Mss., and met an untimely death in the midst of his useful career; our younger son, G. N. Josyer; B.E., who has been helping us in seeing the work through; and our consultants in the course of the work, Sris. Alwar Tirumaliengar and M. A. Tirunarayan, B.E., M.I.E., M. N. Srinivasan, B.Sc, Hons, LL.B., Professor M. A. Tirunarayanan, D.Sc, and Sris M. C. S. Chari, B.Sc., N. Narasimhan, B.E., R. T. Krishnan, B.E., Pandit K. Ramaswamy Iyengar, and Mr. N. N. Sastry of P.T.I., and other associates and assistants.

Sanskrit and English have been our two eyes since some 75 years, and we are placing the unique volume before the elite of the world as an outstanding contribution to world literature from the ever-living past. We hope they will deem it an invaluable addition to their libraries, and find it an ever interesting companion.

The 20th century may be said to be made historic by 2 achievements, the bringing of Moon-rock from outer space, and the publication of "Vymanika Sastra" from the unknown past. The Moon-rock is just rock, not a cluster of shining pebbles from Kimberley of South Africa. But the "Vymanika Sastra" is a Cornucopia of precious formulas for the manufacture of Aeroplanes, which should make Lindbergh, Rolls, Zeppelin, De Havilland, Tupolev, and Harold Gray of Pan American, gape in astonishment, and if duly worked up, herald a new era of Aeroplane manufacture for the benefit of Mankind!

G. R. JOSYER

15-3-1973   Hon. Director, International
Mysore-4, India.   Academy of Sanskrit Research.

MAHARSHI BHARADWAAJA'S

VYMAANIKA-SHAASTRA

OR

SCIENCE OF AERONAUTICS

Part of his unknown work

"YANTRA SARVASVA"

or

"ALL ABOUT MACHINES"

as revealed to venerable
Pandit SUBBARAYA SASTRY
and recorded in hand-written
Sanskrit Manuscript Form

translated into English by
G. R. JOSYER, M.A., Hons., F.R.E.S., M.R.S.L.

Founder Director,
INTERNATIONAL ACADEMY
OF SANSKRIT RESEARCH

Printed at the
CORONATION PRESS, MYSORE 4, INDIA.

Maharshi Bharadwaaja's

VYMAANIKA SHAASTRA

FIRST CHAPTER

Maharshi Bharadwaaja:

I make obeisance to the Divine Being, who is visible on the crest of the Vedas, who is the fountain of eternal bliss, and whose abode is reached by Vimaanas or Aeroplanes. Having studied the Shaastraas or sciences propounded by previous men of science to the best of my ability, for the benefit of mankind, I shall deal with the science of Aeronautics, which is the essence of the Vedas, which will be a source of joy and benefit to humanity, which will facilitate comfortable travel in the sky from world to world, in eight chapters, consisting of 100 topics, in 500 sutras or cryptic pronouncements.

Commentary by Bodhaananda:

I bow to God Mahadeva and His Consort, to Saraswathi Goddess of learning, to Ganapathy guardian of benevolent efforts, and to my venerable preceptor, and I bow to Maharshi Bharadwaaja. In Addition to my own knowledge of Logic, I have five times turned over Vaalmeeki's Mathematics, 'Paribhaashaa Chandrikaa,' and 'Naamaarthhakalpaka,' and aided by their authority, I, Swaamy Bodhaananda, for the easy understanding of the young, have written this 'Bodhananda Vritti,' to elucidate Maharshi Bharadwaaja's concisely worded text on Aeronautics.

At the outset Maharshi Bharadwaaja invokes God in the traditional manner for the successful commencement, progress, and completion of his great literary work. Attaining mastery over the Vedas by Divine Grace, and studying the works of earlier Aachaaryaas or preceptors, he has churned the Vedic lore, and extracting the, cream, presented it to mankind for reaping untold benefits, in the work named 'Yantrasarvasva.' In the fortieth chapter therein he deals with the science of Aeronautics, explaining the construction and use of many kinds of aeroplanes, in 8 chapters, containing 100 subject heads, comprising 500 sutras or oracular pronouncements.

In the first stanza the reference is to the teaching of the sacred works, "Uttara-taapaneeya," 'Shaibya-prasna,' 'Kaataka,' and 'Maandookya,' that the symbolic letter, 'Om,' leads to the knowledge of God and Salvation. Bharadwaaja implies that the Vimaana or aeroplane constructed according to Vymaanika Shaastra, may enable men to reach God, and enjoy the benefits of His Divine abode.

The previous Aachaaryaas Bharadwaaja refers to are named by Vishwanaatha as,—Naaraayana, Shownaka, Garga, Vaachaspathi, Chaakraayani and Dhundinaatha, venerable authors of "Vimaana-Chandrikaa", "Vyomayaana-Tantra," "Yantra-Kalpa", "Yaana- Bindu," "Kheta-yaana Pradeepikaa," and "Vyomayaana-Arkaprakaasha," respectively.

Bharadwaaja thus defines the word Vimaana:

☞ **Vega-Saamyaat Vimaano Andajaanaam. Sootra 1.**

> "Owing to similarity of speed with birds,
> it is named Vimaana."

Bodhaananda Vritti:

The word "andaja" means "egg-born", and includes eagles and other birds which fly by their own volition. The Vimaana is a vehicle which flies in the sky with speed comparable with birds. Lallachaarya says, "That which can fly in the sky with speed equal to that of birds, is called Vimaana."

Aachaarya Naaraayana says, "That which can speed on earth, on water, through air, by its own power, like a bird, is a "Vimaana."

Shankha says, "Experts in the science of aeronautics say, "That which can fly through air from one place to another is a Vimaana"

And Vishwambhara says, "Experts say that that which can fly through air from one country to another country, from one island to another island, and from one world to another world, is a "Vimaana"."

Having thus defined the name of the Vimaana, the sage proceeds to describe its details.

☞ Rahasyagnyodhikaaree. Sootra 2.
"The pilot is one who knows the secrets."

Bodhaananda: Scientists say that there are 32 secrets of the working of the Vimaana. A pilot should acquaint himself thoroughly with them before he can be deemed competent to handle the aeroplane. He must know the structure of the aeroplane, know the means of its take off and ascent to the sky, know how to drive it and how to halt it when necessary, how to manoeuvre it and make it perform spectacular feats in the sky without crashing. Those secrets are given in "Rahasya-Lahari" and other works, by Lalla and other masters, and are described thus:

"The pilot should have had training in maantrica and taantrica, kritaka and antaraalaka, goodha or hidden, drishya and adrishya or seen and unseen, paroksha and aparoksha, contraction and expansion, changing shape, look frightening, look pleasing, become luminous or enveloped in darkness, deluge or pralaya, vimukha, taara, stun by thunderous din, jump, move zig-zag like serpent, chaapala, face all sides, hear distant sounds, take pictures, know enemy manoeuvres, know direction of enemy approach, stabdhaka or paralyse, and karshana or exercise magnetic pull.

These 32 secrets the pilot should learn from competent preceptors, and only such a person is fit to be entrusted with an aeroplane, and not others.

They are explained thus by Siddhanaatha:

1. Maantrika; As prescribed in "Mantraadhikaara," by invoking the mantras of Chhinnamasta, Bhairavee, Veginee, Siddhaamba, acquire the powers of ghutikaa, paadukaa, visible and invisible and other mantraas with potent herbs and efficacious

oils, and Bhuvaneswaree Mantra which confers spiritual and mesmeric powers, to construct aeroplanes, which don't break cannot be cut, cannot be burnt, and cannot be destroyed.

2. Taantrika: By acquiring Mahaamaaya, Shambara, and other taantric powers, to endow the plane with those powers.

3. Kritaka: By study of architects like Vishwakarma, Chhaay-aaparusha, Mann, Maya and others, to construct aeroplanes of various patterns.

4. Antaraala: In the wind-swept atmospheric region of the sky, in the clash at the borders of mighty currents, an inadvertent plane is likely to be smashed to pieces. But by getting warned of the approach of such danger spots, the plane could be halted and steered with care.

5. Goodha: As explained in 'Vaayutatva-Prakarana', by harnessing the powers, Yaasaa, Viyaasaa, Prayaasaa in the 8th atmospheric layer covering the earth, to attract the dark content of the solar ray, and use it to hide the Vimaana from the enemy.

6. Drishya: By collision of the electric power and wind power in the atmosphere, a glow is created, whose reflection is to be caught in the Vishwa-Kriyaa-darapana or mirror at the front of the Vimana, and by its manipulation produce a Maaya- Vimaana or camouflaged Vimana.

7. Adrishya: According to "Shaktitantra", by means of the Vynarathya Vikarana and other powers in the heart centre of the solar mass, attract the force of the ethereal flow in the sky, and mingle it with the balaahaavikarana shakti in the aerial globe, producing thereby a white cover, which will make the Vimana invisible.

8. Paroksha: According to "Meghotpatthi-prakarana," or the science of the birth of clouds, by entering the second of the summer cloud layers, and attracting the power therein with the shaktyaakarshana darpana or force-attraction mirror in the Vi-

mana, and applying it to the parivesha or halo of the Vimaana, a paralysing force is generated, and opposing Vimaanas are paralysed and put out of action.

9. Aparoksha: According to 'Shakti-tantra,' by projection of the Rohinee beam of light, things in front of the Vimaana are made visible.

10. Sankocha, or Contraction: As prescribed in the Yantraangopasamhaara section, when the Vimaana is flying at speed with fully extended wings, and there is danger ahead, turning the 7th switch in the Vimana, its parts can be made to contract.

11. Vistrita: According to 'Akaashatantra', when the Vimana is in the central air flood in the third and first regions of the sky, by turning the switch in the 11th section of plane, it becomes expanded suitably according to "Vaalmeeki Ganita."

12. Viroopa Karana: As stated in "Dhooma Prakarana", by producing the 32nd kind of smoke through the mechanism, and charging it with the light of the heat waves in the sky, and projecting it through the padmaka chakra tube to the bhyravee oil-smeared Vyroopya-darpana at the top of the Vimaana, and whirling with 132nd type of speed, a very fierce and terrifying shape of the Vimana will emerge, causing utter fright to onlookers.

13. Roopaantara: As stated in "Tylaprakarana," by preparing griddhrajihwaa, kumbhinee, and kaakajangha oils and anointing the distorting mirror in the Vimaana with them, applying to it the 19th kind of smoker and charging with the kuntinee shakti in the Vimana, shapes like lion, tiger, rhinoceros, serpent, mountain, river will appear and amaze observers and confuse them.

14. Suroopa: By attracting the 13 kinds of Karaka force mentioned in "Karaka prakarana" applying snow-surcharged air and projecting it through the air conveying tube to the push-pinee-pinjula mirrors in the front right side of the Vimana, and focusing on it the suragha beam, a heavenly damsel bedecked with flowers and jewels will appear to onlookers of theVimana.

15. Jyotirbhaava: As stated in "Amshubodhinee," out of Samgnaana and other 16 digits of the solar glow, by attracting the 12th to the 16th digits and focusing them on the air force in the Mayookha section in the fourth pathway in the sky, and similarly by attracting the force of the etherial glow and mingling it with the glow in the 7th layer of air mass, and then by projecting both these forces through the 5 tubes in the Vimana on to the section of the guhaa-garbha mirror, a rich glow like the morning glow of the sun will be produced.

16. Tamomaya: As described in "Darpana Prakarana," by means of the dark force mirror, capture the force of darkness, pass it through the Thamo-Yantra in the north-west side of the Vimana, and by turning a switch produce at noon-day the utter darkness of the night of the new-moon.

17. Pralaya: As described in the magic book of destruction, attract the 5 kinds of smoke through the tube of the contracting machine in the front part of the Vimana, and merge it in the cloud-smoke mentioned in "Shadgarbha Viveka", and pushing it by electric force through the five-limbed aerial tube, destroy everything as in a cataclysm.

18. Vimukha: As mentioned in "Rig-hridaya", by projecting the force of Kubera, Vimukha and Vyshawaanara poison powder through the third tube of the roudree mirror and turning the switch of the air mechanism, produce wholesale insensibility and coma.

19. Taara: By mixing with etherial force 10 parts of air forte, 7 parts of water force, and 16 parts of solar glow, and projecting it by means of the star-faced mirror through the frontal tube of the Vimana; the appearance of a star-spangled sky is created.

20. Mahaashabda Vimohana: By concentrating the air force in the seven tubes of the Vimana, and turning the switch, produce, as stated in "Shabda prakaashikaa" a crescendo of thunderous din, which stuns people, and makes them quake with fear and become insensible.

21. Langhana: As stated in "Vaayu tattva prakarana" When crossing from one air stream into another, the Vimana faces the baadaba glow of the sun and catches fire. In order to prevent that, the electric force and air force in the Vimana should be conjoined and centred in the life-centre of the Vimana, and by turning the switch, the Vimana will leap into safety.

22. Saarpa-Gamana: By attracting the dandavaktra and other seven forces of air, and joining with solar rays, passing it through the zig-zagging centre of the Vimana, and turning the switch, the Vimana will have a zig-zagging motion like a serpent.

23. Chaapala: On sighting an enemy plane, by turning the switch in the force centre in the middle section of the Vimana, a 4087 revolutions an hour atmospheric wave speed will be generated, and shake up the enemy plane.

24. Sarvatomukha: When a formation of enemy planes comes to attack one's Vimana, by turning the switch at the crown of the Vimana, make it revolve with agility and face all sides.

25. Parashabda Graahaka: As explained in "Sowdaaminee kalaa" or science of electronics, by means of the sound capturing yantra in the Vimana, to hear the talks and sounds in enemy planes flying in the sky.

26. Roopaakarshana: By means of the photographic yantra in the Vimana to obtain a television view of things inside an enemy plane.

27. Kriyaagrahana: By turning the key at the bottom of the Vimana, a white cloth is made to appear. By electrifying the three acids in the north-east part of the Vimana, and subjecting them to the 7 kinds of solar rays, and passing the resultant force into the tube of the Thrisheersha mirror and making the cloth screen face the mirror, and switching on the upper key, all the activities going on down below on the ground, will be projected on the screen.

28. Dikpradarshana: Turning the key at the front of the Vimana the dishaampati yantra will show the direction from which the enemy plane is approaching.

29. Aakaashaakaara: According to "Aakaasha-tantra", by mixing black mica solution with neem and bhoonaaga decoctions and smearing the solution on the outer body of the Vimana made of mica plates, and exposing to solar rays, the plane will look like the sky and become indistinguishable.

30. Jalada roopa: Mixing pomegranate juice, bilva or bael oil, copper-salt, kitchen smoke, granthika or gugul liquid, mustard powder, and fish scale decoctions, and adding sea-shell and rock-salt powder, and collecting smoke of the same solution and spreading it with solar heat enveloping the cover, the Vimana will have the appearance of a cloud.

31. Stabdhaka: By projecting apasmaara poison-fume smoke through the tube on the north side on the Vimana, and discharging it with stambhana-yantra, people in enemy planes will be made unconscious.

32. Karshana: When enemy planes come in strength to destroy one's Vimana, by setting aflame the Jwaalinee shakit in the Vyshwaanara-naala or pipe located at the navel of the plane, and switching the keys of the two wheels to 87 degrees of heat, the burning shakti will envelope the enemy plane and destroy it.

These are the 32 rahasyaas or secrets which should be known by pilots according to Siddhanaatha.

"Maargaadhikaranam " Aerial Routes:

☞ **Panchagnyascha. Sootra 3.**
"The pilot should know five things."

Bodhaananda Vritti:

As the secrets of aeronautics are indicated in the second sutra, the five atmospheric regions are referred to in the third sutra. According to Shownaka, the regions of the sky are five, named, Rekhaapathha, Mandala, Kakshya, Shakti, and Kendra.

In these 5 atmospheric regions, there are 5,19,800 air ways traversed by Vimanas of the Seven Lokas or worlds, known as Bhooloka, Bhuvarloka, Suvarloka, Maholoka, Janoloka, Tapoloka and Satyaloka.

Dhundinaatha and "Vaalmeeki Ganita" state that Rekha has 7,03,00,800 air routes, Mandala has 20,08,00200 air routes, Kakshya has 2,09,00,300 air routes, Shakti has 10,01,300 air routes, and Kendra has 30,08,200 air routes.

According to "Vaalmeeki Ganita" in the Rekhaapathha region, sections 1 to 4 are suitable for the passage of the Vimanas of this Bhooloka. In the Mandala region sections 3 to 5 are suitable for Vimanas of Bhuvarloka, Suvarloka, and Maholoka dwellers. For the Janoloka Vimanas sections 2 to 5 in the Kakshya region are suitable. Section 1 to 6 in the Shakti region are suitable for the Vimanas of Tapoloka. For the dwellers of Bramhaloka sections 3 to 11 in the Kendra region are suitable, according to shaastras like "Vaalmeeki Ganita" and others.

Maharashi Bharadwaaja:

☞ **Aavartaascha. Sootra 4.**
"Whirl-Pools"

Aavartaas or aerial whirlpools are innumerable in the above regions. Of them the whirlpools in the routes of Vimanas are five. In the Rekhapathha there occurs "Shaktyaavarta" or whirlpool of energy. In Mandala-pathha there occurs the whirlpool of winds. In Kakshyaa-pathha there occurs Kiranaavarta o whirlpool from solar rays. In Shakti-pathha there occurs shytyaavarta or whirlpool of cold-currents. And in Kendra-pathha there occurs gharshanaavartha or whirl-pool by collision. Such whirlpools are destructive of Vimanas, and have to be guarded against.

The pilot should know these five sources of danger, and learn to steer clear of them to safety.

Maharshi Bharadwaaja:

☞ **Angaanyekatrimsat. Sootra 5.**
"The parts are thirty one"

Bodhaananda Vritti:

Just as the human body, if it is complete in all its limbs, is best able to accomplish things, the Vimana, if it is complete in all its parts, will be capable of functioning efficiently. From the location of the Vishwakriyaadarpana 31 locations of Vimana components are mentioned.
According to "Chaayaapurusha Shaastra" they are:

1. Vishwakriyaadarpana or mirror of outside views.
2. Shaktyaakarshana or energy attracting mirror.
3. Parivesha mechanism above the hood of the Vimana.
4. Angopasamhaara yantra or folding up yantra at the 7th bindukeelaka.
5. Vistritakriyaa or opening out yantra location in the middle of the 11th section.
6. Vyroopya darpana and
7. Padmachakramukha at the shirobhaaga or crest of the Vimana.
8. The Kuntinee-shakti mechanism is to be in the neck of the Vimana.
9. Pushpinee and Pinjulaa Mirrors are to be in the right side of the centre.
10. At the front of the left side are to be located the Naala-panchaka or 5 pipes.
11. Guhaagarbha mirror yantra is to be in the front part of the stomach of the plane.
12. Thamoyantra at the north western side.
13. Pancha-vaataskandha-naala on the western centre.
14. Rowdree mirror.
15. Vaataskandha keelaka at the bottom centre.
16. Shaktisthaana at the front and right sides.
17. Shabda-kendra-mukha at the left side.
18. Vidyuddwaadashaka at the north-east side.
19. Praanakundala at the moola of the Vimana.

20. Shaktyudgama at the navel of the Vimana.
21. Vakraprasaarana at the side of Vimanaadhaara.
22. Shaktipanjara in the central portion.
23. Shirahkeelaka at the head of the Vimana.
24. Shabdaakarshaka yantra at the shoulder.
25. Pata-prasaarana at the bottom centre.
26. Dishaampati yantra at the left front.
27. Pattikaabhraka at the centre of the hood of the Vimana.
28. Solar power attractor at the top of the Vimana.
29. Apasmaara or poison gas at the sandhi-naala mukha or junction tube front.
30. Sthambhana yantra at the bottom.
31. Vyshwaanara-naala at the navel centre.

This is the placing of the 31 components of the Vimana.

"Vasthraadhikaranam ": On Clothing

Maharashi Bharadwaaja:

☞ **"Yantu-praavaraneeyow prithak prithak ritubhedaat." Sootra 6.**
"The clothing should be different for different seasons."

The sootra defines the clothing which is to be provided to the pilot in accordance with different seasonal conditions. The impact of the sun's myriad rays on the revolving earth causes seasonal climatic changes. Their effects on human life are either wholesome or unwholesome, as the case may be. The latter cause cramps, drain blood, and denude the body of fat, flesh, and other ingredients. The evil forces of the seasons are reckoned as 25, and affect the skin, bone, flesh, fat, muscles, nerves, joints and other parts of the pilots' body. The clothing provided to them should be such as to safeguard against such effects, and maintain their efficiency.

According to "Pata-samskaara Ratnaakara", silk, cotton, moss, hair, mica, leather, are to be purified by 25 processes, washed with mica-saturated water, and spun into yarn as prescribed by Gaalava. Then fibres from the ketaki flower palm, arka or swallow wort or madar, sun flower tree, cocoanut and jute, should each be purified 8 times as prescribed and by 19 processes, spun into yarn, and woven into cloth. Then the cloth should be soaked in the oils of linseed, tulasi or basil, gooseberry, shamee or acacia suma, bael, and mustard, and dried in the sun 5 times daily for 7 days. Then yellow ochre, lac, tamarind, honey and gingelly manure and mica in equal parts and yena-kshaara salt, put in a crucible vessel, placed in koorma furnace, and boiled with the aid of 3 faced bellows. 8 seers of linseed juice should be added to it. Bees wax, mica, shinjeera, vajra, borax, and ashoka fruit should be boiled, and their oil mixed with the other composition, and boiled in garbhataapana yantra. Then the cloth should be soaked in that decoction and dried 5 times. With this material, fashioning the apparel and clothes of the pilots handsomely, according to the types of the cloth and requirements the crew, as prescribed by Agnimitra, and handing it to them to wear, they should be conferred benediction, given a protective amulet and then sent out with cheers. It will ward off evils, promote fitness of body and health of mind, and improve their strength, energy, and competence.

Aahaaraadhikaranam: On Food

Maharshi Bharadwaaja:

☞ "Aahaarah Kalpabhedaat" Sootra 7.
"Food according to Seasons."

Bodhaananda Vritti:

As stated in Kalpa Sootra, the food of the pilots is of three kinds, according to the seasons. "Ashana-Kalpa" or "Principles of Diet", says—"During the spring and summer months, the pilot's food should consist of buffalo-milk among liquids, among grains aadhaka or tuvar-dhal, and among flesh, the flesh of sheep.

In the 4 months of rains and autumn, cow's milk among liquids, wheat and black-gram among grains, and flesh of cocks and hens.

In the 4 months of winter and snow, goat's-milk, yava and black-gram among grains, and flesh of sparrows.

For pilots belonging to the three Dwija castes of Braahmin, Kshatriya and Vyshya, the food will not include flesh.

Maharshi Bharadwaaja:

☞ "Visha Naashas Tribhyaha" Sootra 8.
"The three Varieties ward off seasonal evil effects"

Bodhaananda Vritti:

The 25 kinds of poisonous effects of the seasons are warded off by alteration in the food so as to suit the seasonal conditions.

"Vishanirnaya-adhikara" states,—

The seasons are each differently conditioned by the changes in the watery forces in the sky. The 101 forces in the aerial atmosphere, colliding with the 1/16th force in the watery sky in the seventh region, at the sineevaalee and kuhoo yogas or full-moon and new-moon conjunctions, produce maleficent and beneficent effects. The beneficent forces are 7,58,00,700 in number, and the maleficent forces also are of the same number, according to "Vaalmeeki Ganita". The beneficent effects are during the full moon period, and the maleficent effects during the new-moon period. 25 maleficent poisonous forces known as Bhedinee, tend to paralyse the pilots' physical effeciency. That is avoided by altering their food according to seasons. So says Sage Shaataatapa.

By such adjustment the pilots' physical fitness will be maintained.

Maharashi Bharadwaaja:

☞ "Tat Kaalaanusaaraat iti" Sootra 9.
"That at set times"

Bodhaananda Vritti:

Having defined the seasonal types of food, the meal times are now defined. According to Shownaka, the times for taking meals are prescribed as follows: Family men should take food twice a day, or once a day. Ascetics should take food once a day. Others can take food four times a day. Air pilots should take food 5 times a day. And yogis may take as many times as they like.

According to "Lalla-kaarika" or "Diet Rules" by Lalla,

Food should be taken at the end of the 2nd yaama-(yaama=3 hours) in the day time, and at the end of the 1st yaama in the night. That is for family men. If they take only one meal a day, it should be during the 3rd and 4th yaamas. Sanyaasis or ascetics who eat only once a day should dine as above. For the labouring classes the times are thrice during day time, and once in the night. For pilots of Vimanas the meals are thrice during day time, and twice at night.

Maharshi Bharadwaaja:

☞ "Tadabhaavay Sathva-Golovaa" Sootra 10.
"If unavailable, then vitamin pills or food-balls"

Bodhaananda Vritti:

If the prescribed food-stuffs are not available for use during their flights, then essence extracts made by proper cooking with admixture of spices and condiments into potable and eatable form, or food balls-made out of them should be supplied to the pilots for consumption on flights.

Says "Ashana kalpa" or "Food manual",

"There are 5 kinds of food, that are nutritious and whole-some; cooked rice or grain, gruel, cooked flour, baked flat bread, and preparations made out of essence-extracts from food mate-rials. The last named are superior to all the others".

"Paakasarvasva" or "Art of cooking" observes,
"Removing the husk and other non-food parts from it by machines, the grain should be made into flour and cooked in a suitable vessel and when it has reached the 8th degree of reduc-tion, add essences, sweets, condiments, and ghee, and prepare food-balls, having nice flavour and delicious taste, and being nourishing to the body.

Maharshi Bharadwaaja:

☞ **"Phala Moola Kanda Saarovaa." Sootra 11.**
"Or essence of fruits, roots, and bulbs."

Bodhaananda Vritti:

In this sootra it is stated that preparations made from edible roots, potato and other bulbous vegetables, and from fruits are also suitable as food.

"Ashana-Kalpa" says,

If food made of grains is not available, that from roots, bulbs, and fruits may by used as food, in the form of flour, sug-arcandy, manjoosha or jaggery, honey, milk, ghee, oily-products, and roots and berries whichcontain sweet, salt, pungent, acrid, and alkaline tastes. Such roots are said to be 56 in number. They should be purified, powdered, and duly cooked, and made into balls, and given out for use as food.

Similarly the bulbous vegetables which are of 16 kinds, and fruits which are of 32 kinds, and food prepared out of them are excellent food, Food from roots develops brain, nourishes the body, strengthens the bones, and gives virility. Food from bulbs promotes brilliance, and bodily vigour, and strengthens the life current. Food from fruits nourishes mind, intelligence, blood, flesh, and vital liquids. Therefore these alternatives are recommended for pilots of Vimanas.

Maharshi Bharadwaaja:

☞ **"Apicha Trinaadeenam." Sootra 12.**
"Even grasses, herbs and shrubs."

Bodhaananda Vritti:

This sootra indicates that even grasses, herbage, and creepers, could be made to yield food.

Says "Ashana-Kalpa",

Like roots, bulbs and fruits, grasses, shrubs and herbs, provide good food for men. Six kinds of doorva grass, 6 kinds of munja hemp, 6 kinds of darbha or long grass, 6 kinds of shoundeera, and 6 kinds of Ashwakarna or sal, or mimordica charantia, Shatamoolee of 3 kinds, Kaaruvellee; Chandravellee, Madhuvellee, Varchulee, Makutee vellee, sugandhaa, and sooryavellee may be made to yield good food, nutritious and bracing.

Selected by men who know them well, these vegetation, including their flowers, shoots, and leaves, by proper cleaning and cooking, may be made to yield solid or liquid food, which will serve as satisfactory substitute food for pilots of Vimanas. And Somavallee or moon-plant, Chakrikaa, Rasavallikaa, Kooshmandavallee, Ikshuvallee, Pishtavallaree, Sooryakaanta, Chandrakaanta, Meghanaada, Punarnava, Avantee, Vaastu, Matsyaakshee, and Rukma and others, provide good bases for lasting food, duly mixed with sweets and condiments.

Lohaadhikaranam: Metals

Maharshi Bharadwaaja:

"Athha Yaana Lohaani." Sootra 13.
"Next, the Metals for aeroplanes"

Bodhaananda Vritti:

Having dealt with clothing and food for pilots, now the metals suitable for aeroplanes are being dealt with.

Says Shounaka:

There are 3 kinds of metals named somaka, soundaalika, and mourthwika. By mixing them, 16 kinds of heat-absorbing metals are produced.

Their names are ushnambhara, ushnapaa, ushnahana, raajaamlatrit, veerahaa, panchaghna, agnitrit, bhaarahana, sheetahana, garalaghna, amlahana, vishambhara, vishalyakrit, vijamitra and Vaatamitra etc.

"Maanibhadra Kaarika," or "Dictas of Manibhadra," Says, "Metals which are light, and are suitable for producing aeroplanes are 16. They are heat absorbing, and should be used in the manufacture of aeroplanes."

Saamba also says that the 16 metals formed by mixing the root metals, soma, soundaala and mourthwika, are non-heat-conductors and are useful for Vimanas. Their characteristics are now examined.

In the 7th layer of the earth, in the third mine therein, metals of the Soma series are found. They are of 38 kinds. Among them there are three from which Ooshmalohas or heat resisting metals are to be extracted. "Lohatantra" or "Science of Metals" also says that in the 3rd section of the 7th layer of the earth, Metals of Souma class, possessed of 5 special qualities, are called "beejalohas" or "root- metals".

There are 3000 metal bearing layers within the earth. Of them 1300 layers contain the better quality. In the 7th layer metals are of 27 types. The 3rd type of metals are of five-fold qualities, and are known as root metals. The origin of metals of the Soma class is thus described in "Lohakalpa.":

"The gravity of the centre of the earth, the gravity of global earth, the solar flood, the air force, the force emanating from the planets and stars, the sun's and moon's gravitational forces, and the gravitational force of the Universe, all together enter the layers of the earth in the proportion of 3, 8, 11, 5, 2, 6, 4, 9, and, aided by the heat and moisture therein, cause the origin of metals, of various varieties, grades and qualities."

The Souma group of metals are named, as per sage Atri, in "Naamaartha Kalpa". "Souma, Sowmyaka, Soundaasya, Soma, Panchaanana, Praanana, Shankha, Kapila are the names of the Souma metals, with distinct qualities indicated by their names."

The name "Souma" consists of sounds, s, on, ma, and ha, "Paribhaasha Chandrika" and "Vishwambhara Kaarikaa" state, "The oceanic force and solar force instil 4 kinds of forces into root metals. The sum total of the forces are said, according to "Vaalmeeki Ganitha" to number 1, 67, 768. Some of these forces are indicated by the sound "s" Some of the forces emanating from the sun and the elements are indicated by the sound "ou". Similarly other concerned forces are indicated by the letters "ma" and "ha".

The Varuna and Soorya force contents of all root metals are of four groups. In each group the force content is said to be 1, 67, 768. Of the Koorma and Kashyapa forces of Vaaruna group, the 67th from Ooshaa koorma, and the 85th Kaashyapa force, called "Kaala", are indicated by the letter "Sa".

Of the solar group of forces, maartaanda and bhoota 71st, and the ruchika force 160 are indicated by the sound "ra". Similarly, of the forces of sun and stars in aditi, the 9th called "Sundaa", and the stellar force 101 called "Bhowma" are indicated by the letter "Ma". And in the dhruva varga, soma and baadaba forces, 109 and 14 respectively, are indicated by the visarga sound "ha".

The four forces working inside the earth, by flux of time mature into the Souma type metal.

In Soundaala metal, the 11th force, dhanadaa, in Koorma is indicated by letter "sa". The 110th Kaashyapee force, rook, is indicated by the sound "ow". The sun's 100 powered dravamukhee shakti, and bhoota-shakti known as anvee 700 powered, are together indicated by the anuswaara sound "m". The sun's kaantaa shakti 49, and the stars' 25 shaktis, varchaa, are indicated by the letter "da". Similarly the soma forces in dhruva varga, is indicated by the long "aa" in "daa". The moon's 364 ujwalaa and baadabaa's 500 known as kaala are indicated by the letter "la".

That is "Soundaala".

Regarding the third, "Mourtwika", Koormashakti, paarthiva 1300, is indicated by the letter "ma". Kaasyapa shakti, kaalima 2001, is indicated by the sound "ow". Maartaanda shakti, laaghava 260, is indicated by the sound "r": bhootashakti, vaarchulee 37, by the letter "tha": stellar force, rukshmaka 1063, is indicated by the letter "va". Arkashakti, varuna 113, is indicated by the sound "e": soma force rijukaa 8009, and pooshnikaa 1012, are indicated. by the letter, "ka".

Lohashuddhyadhikaranam: Purification of Metals.

Maharshi Bharadwaaja:

☞ **"Tatchhuddhir yathaa shodhanaadhikaaray." Sootra 14.**
  "Their purification is as per shodhanaadhikaara".

Bodhaananda Vritti:

Taking soma metal first, it should be filled in a wide-brimmed
vessel and adding jambeera or citron juice, likucha or lime juice,
vyaaghra or castor, chinchaa or tamarind, and jamboo or rose
apple juices, it should be boiled to 27th degree of heat for a day.
Then taking it out and washing it, it should be boiled in 5 kinds
of oils, 4 kinds of acids, and 7 kinds of decoctions.

They are named in "Samskaara Darpana":

Gunjaa or wild liquorice, Kanjala, castor, kunjara, and karanja or Indian beach oils, praana-kshara, viranchi, kanchuki, and
khura acids, and hingoo or asafoetida, parpata, ghontikaaa,
jataa-maamsee or spikenard, white gourd or Vidaaraanginee,
and matsyaakshee decoctions.

That is the process of purification of soma metal.

The purification of Soundaala metal is like that of soma with
regard to boiling in the cauldron, but the process of purification
is with 6 acids, 7 oils, and 5 decoctions. They are, according to
Samskaara Darpana."—

Ingaala or, ingudee, gouree or reddish herb, couries, grapes,
rata, aapya, and ulbana oils, ankola, mushti, shankha, bhallaata-
ka, kaakola, and virancha acids, and kuluththa or horsegram,
nishpaava, sarshapa or mustard, aadhaka, and wheat decoctions
or gruels.

Mourthweeka metal also should be baked like soundaala,
and then should be boiled with shivaari oil, kudupa acid and
vishambharee leather decoction.

Having defined the root-metals and their purification, we
next consider the casting of Ooshmapaa loha.

SECOND CHAPTER

Maharshi Bharadwaaja:

☞ **"Oosh mapaastriloha Mayaaha." Sootra 1.**
"Ooshmapaa metals are made up of 3 metals."

Bodhaananda Vritti:

The heat-proof metals are made out of the three, Souma, Soundaala, and Morthweeka mentioned in the previous chapter. It is said in "Loha Ratnaakara" that each of the three yields varieties of seed metals. Their names are, in souma group,—souma, soumyaka, sundaasya, soma, panchaanana, ooshmapa, shaktigarbha, jaangalika, praanana, shankha, and laaghava; The names of the metals of soundeera origin, are viranchi, sourya-pa, shanku, ushna, soorana, shinjikaa, kanku, ranjika, soundeera, mugdha, and ghundaaraka. In the mourthweeka group, the 11 are anuka, dvyanuka, kanka, tryanuka, shvetaambara, mridam-bara, baalagarbha, kuvarcha, kantaka, kshvinka and laghvika.

Maharshi Bharadwaaja:

☞ **"Melanaath" Sootra 2.**
"By Mixing"

Bodhaananda Vritti:

The said metals are to be mixed in requisite proportions and melted. It is said in "Lohatantra" that ushnambhara metal is pro-duced by mixing numbers 10, 5, 8 of soma, soundala, mourth-weeka groups of metals respectively in the proportion of 1, 3, 7, and mixing with one third the quantity of tankana or borax and melting in the crucible. Similarly taking metals no, 3, 5, and 7 respectively in the three groups in the proportion of 4, 1, and

8, and mixing with tankana, and melting in crucible, the metal ooshma is obtained. Metal ooshmahana is produced by melting metals 2, 5, and 9 from the three groups in the proportion of 6, 3, and 7, with tankana. Metal Raaja is produced by melting nos. 3, 8, and 2 of the three groups as before. Similarly metal Aamlatrit is produced by taking numbers 9, 7, 1 in the three metal groups, in the proportion of 10, 7, 8 and mixing with tankana and melting as prescribed.

Similarly metals 6, 4, 5, respectively in the proportion of 5, 5, 12, melted with tankana or borax, will yield the metal veerahaa. The metal panchaghna is got by taking numbers 8, 6, and 4 of the three groups in the proportion of 20, 18, 26, and mixing with tankana or borax and melting.

The metal agnitrit is produced by mixing numbers 5, 2, 10, in the proportion of 30, 20, and 10, and melting with borax in the crucible. The metal bhaarahana is produced by mixing numbers 7, 11, and 6 in the three groups in the proportion of 5, 12, and 7, mixing with borax, and melting in the crucible.

To produce metal sheetahana, metals 10, 9, and 3 in the three groups respectively, in the proportion of 22, 8, and 10, should be mixed with borax and melted in the crucible. Garalaghna is produced by taking numbers 11, 10, and 11 in the three groups in the proportion of 20, 30, and 8, and melting with borax in the crucible.

Similarly Aamlahana is produced by taking numbers 11, 8, and 4 in the three groups in the proportion of 20, 12, 36, and melting with borax in the crucible. Metal Vishambhara is produced by taking numbers 19, 8, and 10 in the three Ooshmapa groups respectively in the proportion of 20, 12, and 6, and melting with borax in the crucible.

Metal vishalyakrit is produced by taking numbers 3, 5, and 11, in the proportion of 20, 12, and 6, and melting in the crucible with borax. Dwijamitra is produced by taking numbers 8, 3, and 9 in the proportion 5, 8, 10, and melting with borax in the crucible. And metal Vaatamitra is produced by taking numbers 8, 6, and 5 in the three groups of Ooshmapa metals, in the proportion of 22, 8, and 10, and adding borax and melting in the crucible.

Mooshaadhikaranam: The Crucible.

Maharshi Bharadwaaja:

☞ **"Panchamaad dwitheeyay " Sootra 3.**
"From the 5th variety in the 2nd group."

Bodhaananda Vritti:

According to "Nirnayaadhikaara", the melting of the superior, medium, and inferior kinds of metals is to be done in 407 different kinds of crucibles. They are divided into 12 groups. For the melting of the root-metals the second group of crucibles is considered the best.

Lallacharya also states that metallurgists mention 12 kinds of metals: kritaka or artificial, apabhramshaka or corrupted, sthalaja or mud-born, khanija or found in mines, jalaja or aquatic, dhaatuja or mineral-born, oshadhivargaja or vegetation-born, krimija or evolved from vermin, maamsaja or flesh-born, kshaaraja or grown from salts, baalaja or hairborn, and andaja or resultant from eggs. Different classes of crucibles are to be used for melting different kinds of metals. In the second class of crucibles there are said to be 40 varieties. Of them, number 5, known as antarmukha or inward-mouthed, is prescribed for melting the root-metals.

It is described in "Mooshaakalpa" or art of making crucibles. 8 parts of gingelly manure or black-gram flour, 4 parts of metal rust, 3 parts of metal, 3 parts of laangalee or jussieuea repens or gloriosa superba, 6 parts of gum arabic, 2 parts of ruruka, 3 parts of salt- petre, 5 parts of creepers, 6 parts of charcoal, 5 parts of 5 kinds of grasses, 4 parts of paddy husk ashes, 2 parts of red arsenic, 2 parts of naagakesara, 5 parts of varolika flower,

5 parts of borax, 2 parts of black laamancha or scented grass or andropogon muricatus, 5 parts of sindoora or red ochre, 2 parts of gunja seeds or wild liquorice, 4 parts of sea-foam, all these are to be ground and made into fine flour, to which are added equal quantity of gum and 5 parts of earth and dust, and the whole is baked in a vessel with shivaaree oil for 3 yaamaas or 9 hours. When the contents have unified and become properly fluid, it should be poured through the nozzle into the crucible mould, and allowed to rest. The resultant crucible, known as "antarmukha," would be best suited to melt the metals required for producing a Vimaana.

Athha Vyaasatikaadhikaranam: The Fire-place.

Maharshi Bharadwaaja:

☞ **"Athha Kundas-Saptamay-Nava". Sootra 4.**
"Then fire-place, number 9 in class 7."

Bodhaananda Vritti:

Having dealt with crucible in the last sootra, we now consider the fire-place.

Experts mention 532 varieties of vyaasatikaas or fire-places. Of them Koorma-vyaasatika, or tortoise-shaped fire-place is best suited for melting the seed-metals for the Vimaana.

Kunda-kalpa or the art of furnace construction mentions 532 kinds of furnaces. They are divided into seven classes, each including 76 varieties. Furnace no. 9 in the 7th class, is best suited for melting the requisite metals of the vimaana, and its name is koorma- vyaasatikaa, or tortoise-shaped furnace.

It is said in "Kunda-nirnaya ", that on a prepared ground, a quadrilateral or circular shaped furnace 10 feet wide should be constructed, shaped like a tortoise. In order to place the bellows, there should be constructed a pedestal shaped like a tortoise, and with five faces. In the middle of the furnace arrangements should be made for placing the crucible. On either side of the furnace there should be an enclosure for stocking charcoal. And on either side there should be a mechanism for receiving the molten metal.

Athha Bhastrikaadhikaranam: The Bellows.

Maharshi Bharadwaaja:

☞ **"Syaad-bhastrikaashtame Shodashee" Sootra 5.**
"The bellows should be number 16 in the 8th class."

Bodhaananda Vritti:

The making of bellows is referred to in this sootra. It is said in "Bhastrikaa Nibandhana," that as there are 532 kinds of furnaces, there are 532 kinds of bellows. Narayana also says that there are 532 varieties of bellows used in melting metals. They are divided into 8 classes. In the eighth class, the variety numbered 16 is the one suited for the tortoise—shaped furnace. The construction of bellows is described in the work "Bhastrikaa-Nibandhana", as follows:

The barks of suitable trees, leather, thick cloth made from milk cream, bark of areca-nut palm tree, and trinetra (bael? Bengal Quince?), shundeera, suranji, silk-cotton, sheneera, munjaakara, and jute by due processing yield suitable cloth of 605 varieties with which pretty and attractive bellows could be made, with fittings of wood or copper.

THIRD CHAPTER

Darpanaadhikaranam: Mirrors & Lenses.

Maharshi Bharadwaaja:

☞ **"Darpanaashcha" Sootra 1.**
"Mirrors".

Bodhaananda Vritti:

This chapter deals with the mirrors and lenses which are required to be installed in the vimaana. They are seven different ones. Their names are given by Lalla in "Mukura-kalpa" as Vishwakriyaadarpana, or television mirror, Shaktyaakarshana darpana or power-capturing mirror, Vyroopya darpana or appearance changing mirror, Kuntinee darpana, Pinjulaa darpana, Guhaagarbha darpana, and Rowdree darpana or terrifying darpana.

Vishwakriyaa darpana is to be fixed on a revolving stand near the pilot so that he could observe whatever is happening outside on all sides. Its manufacture is thus described in Kriyaasaara:

Two parts of satva, 2 parts of shundilaka, one part of eagle bone, 5 parts of mercury, 2 parts of the foot-nails of sinchoranee, 6 parts of mica, 5 parts of red lead, 8 parts of pearl dust, 18 parts of the eyeballs of sowmyaka fish, one part burning coal, 8 parts of snake's slough, 3 parts of eye pigment, 6 parts of maatrunna, 10 parts of granite sand, 8 parts of salts, 4 of lead, 2 parts of sea foam, 3 parts of white throated eagle's skin, 7 parts of bamboo salt, 5 parts of vyraajya or white keg tree bark, these ingredients should be purified, and weighed, and filled in a beaked crucible and placed in the furnace called chandodara and subjected to a 800 degree heat, and when duly liquified, should be poured into the funnel of the kara-darpana yantra or hand-mirror mould. The result will be an excellent mirror in which will be reproduced minute details of the phenomena outside.

Next Shaktyaakarshana darpana:

As the vimaana flies through the regions of the sky, three classes of destructive forces tend to overcome it. This mirror is capable of neutralising and overcoming their effects.

Dhundinaathaachaarya also says: The wind, solar rays, and fire are known as trivargas. Each of the three has 122 evil effects on the plane's pilot. Those evil forces this mirror will absorb and nullify.

Paraankusha also says: There are certain crucial regions in the air routes of the vimaana, at which the wind, solar heat and fire have 366 malefic influences, and shaktyaakarshana mirror is meant to safeguard against them. It is to be prepared as follows:

5 parts of haritaala or yellow orpiment, 5 parts of virinchi, 8 parts of salts, 4 parts of gingelly husk, 6 parts of diamond, 1 of red mica, 8 parts of burning coal, 3 parts of sand, 2 parts of tortoise egg, 3 parts of bhaarani, 3 parts of kanda, 5 parts of powshkala, 5 of coral, 2 of pearl, 6 of sea-shell, 8 of borax, 3 parts of Bengal quince seed, and 5 of shankha or conch, cleaned, powdered, filled in swan crucible, placed in mandooka furnace and boiled to 500 degrees and poured slowly into vistritaa-mirror yantra will yield a fine shaktyaakarshana mirror.

The Vyroopya—darpana Mirror:

When enemy planes with men intent on intercepting and destroying your vimaana attack you with all the means at their disposal, the viroopya-darpana will frighten them into retreat or render them unconscious and leave you free to destroy or rout them. The darpana, like a magician, will change the appearance of your vimaana into such frightening shapes that the attacker will be dismayed or paralysed. There are 27 such different shapes that are said to be possible. Sammohana-kriyaa-kaanda, or the work dealing with the methods causing insensibility, mentions 17 of them. They are fire, water, wind, thunder, lightning, fumes, scorpion, bear, lion, tiger, and giant-sized frightful birds.

The manufacture of this mirror is thus given in Darpana-pra-karana:

5 parts of bone salt, 3 parts zinc, 3 of lac, 8 of iron, 3 parts of shashabola, 2 of raajakurantika, 8 parts of charcoal ashes, 3 of borax, 8 parts of nakhaa, 7 of sand, 6 of matrunna, 2 of sun-crystal, 3 parts of poora or lime, 25 of mercury, 3 of yellow orpiment, 4 parts of silver, 6 of kravyaada, 8 of garada, 3 of pishta, 4 parts of arshoghna root, 3 parts of vaaraaha pittha, 3 of ammonium chloride, 25 of liquorice oil, taking these and 7 times purifying, filling in crucible, and placing in furnace and boiling it to 800 degrees and pouring into Darpanaasya yantra, will yield an excellent Vyraajaka mirror.

The Kuntinee Mirror:

We now consider the Kuntinee mirror. The wise say that the mirror by the glare of whose rays people's minds get deranged is Kuntinee mirror. Paraankusha says that in the region of the solar electric heat waves of the sky, seven streams of poisonous whirl- winds derange the mind. Scientists have discovered the Kuntinee mirror as a protection against that evil effect.

In "Sammohana-kriyaa-kaanda," the evil forces are described as follows:

Fat, blood, flesh, marrow, bone, skin, intelligence are adversely affected by the evil wind currents known as gaalinee, kuntinee, kaalee, pinjulaa, ulbanaa, maraa, in the electric heat wave regions of the upper sky.

The manufacture of this mirror is thus explained in "Darpana-prakarana":

5 parts of sowraashtra earth, 7 parts of snake's slough, 3 of sea-foam, 5 of shanmukha seeds, 8 of zinc, 3 parts of rhinoceros' nails, 8 of salts, 7 of sand, 8 of mercury, 4 of conch, 6 parts of matrunna, 3 parts of yellow orpiment, 4 of elephant and camel salts, 7 parts of suranghrikaa, 5 of gingelly oil, 8 of pearl-shells, 3 of sea-shells, 4 parts of camphor, purified and filled in shinjikaa crucible, and placed in shinjeera furnace and boiled to 700 degrees, the fluid poured into the Darpanaasya yantra, will form into a morning sun-like kuntinee mirror.

The Pinjulaa mirror:

The conflicting inter-action of the solar rays is called pinjulaa. It has deleterious effect on the black eye-balls of the pilots. The pinjulaa mirror, by intervening will prevent the eye-balls being blinded by the evil rays.

It is said in "Amsubodhinee", or the work on solar rays, "There are four directions, east, west, north and south, and four corner directions, south-east, south-west, north-east and north-west. The solar force of each direction has got its own intensity, owing to different fire-force, different seasonal force, the effect of the five winds, combined with the vaarunee or liquid force of the clouds, and the resulting tension gives rise to four evil forces, andha, andhakaara, pinjoosha, and taarapaa, whose glows, known as rakta, jaathara, taaraagra, and prabha, striking the eye-balls result in blindness of both eyes. "

The manufacture of this mirror is thus described in "Darpana prakarana":

6 parts of goat's milk, 5 parts of red-lead, 8 parts of salts, 7 parts of sand, 5 parts of tree-gum, 8 parts of borax, 2 parts of dambholi essence, 8 parts of mercury, 2 parts of copper and 2 of lead, 4 parts of surolika essence, 8 parts of twak, 3 parts of vaardhyushika, 3 of kanda, 4 parts of pishta or gingelly husk, 3 parts of orpiment, 7 parts of Tinnevelli senna, 4 parts of vrikodaree seeds, these 18 to be purified, powdered, and filled in crucible, and placed in furnace and boiled to 700 degrees, and poured into Darpana yantra, will yield an excellent pinjulaa mirror.

Next Guhaa-garbha darpana:

"The conflict between the electricity in the clouds, wind, and rays, generates forces harmful to pilots. The guhaa-garbha darpana, by attracting them and projecting them by electric force against enemy planes, renders the persons inside them physically disabled and incapable of fighting."

"Prapanchasaara" also says:

"In the Middle of the two shells above kashyapa, there is vaarunee force. Between the shell and vaarunee force 5000 wind currents subsist. Similarly there are disease causing rays numbering 80 millions. The various winds and rays by mutual action result in flows and counter flows. When the cloud force, wind force, and solar force interplay they give rise to various harmful forces like bubbles:"

Lallaacharya also avers, "In accordance with the 110th principle, when the cloud-power, wind-power and sun-power meet with force and collide, they produce poisonous effects which are dangerous to mankind."

Vasishtha says in "Swatassiddha-Nyaaya" or "self-evident truth", that when alien forces cross one another, a poisonous flow will result naturally as an egg comes out of a tortoise.

"Sammohana kriyaa-kaanda" explains:

"By the conflict of cloud force, wind force and solar forces, 305105 poisonous waves known as guha and others emanate, and cause, kushtha, apasmaara, grihinee, khaasa, and shoola. Chief among them are five, known as gridhnee, godhaa, kunjaa, roudree, and guhaa. By accelerating them and directing them against the enemy, the guhaa-garbha mirror disables them."

"Darpana prakarana " describes its manufacture thus:

7 parts of couries, 3 parts of manjula or madder root, 6 parts of sea-foam, 8 parts of ranjaka or phosphorous, 6 parts of mandoora or rust, 8 parts of mercury, 3 parts of orpiment, 7 parts of brahmika, 2 parts of lead, 8 parts of eye pigment, 6 parts of matrunnna, 8 parts of sand, 6 parts of kishora, 5 of muchukunda, 2 parts of gingelly oil, 25 of lohika, 5 parts of mridaani garbha essence, 8 of sowraashtra earth, 5 parts of sphatika, 3 of bones, 15 of indusatva or moonstone?, and 5 of dambholi taakaa dwaya satva, taking these 22, purifying and powdering them and filling crucible, and placing in furnace and boiling to 700 degrees, and cooled in yantra, guhaa garbha darpana is produced.

Rowdree-darpana is a mirror or lens which liquefies everything that it flashes against.

Paraankusha says that where Rudraanyosharaa and abhral-inga come into contact, a fierce force called roudree comes into being. Mingling with solar rays it melts everything. "Sammohana kriyaa kaanda" says:

"By the mixing of roudree and solar rays an evil force called maarikaa is generated, and impelled by the solar electricity, it destroys the enemy planes."

Darpanaa prakarana describes its manufacture:

8 parts of lead, 3 parts of shaalmali, 7 of durvaara, 8 parts kudupinjara, 21 parts of droonee, 8 parts sun-crystal, 27 parts of rudraanee-graavoshara, 6 parts betel leaves, 8 parts of kow-tila, 30 of veeraabhra linga, 8 parts of salts, 7 of sand, 6 parts of matrunna, 3 of dimbhika, 8 of zinc, 13 of ant-hill earth, 6 of gum, 3 of kumbhinee, 3 parts sweet oil, 27 of Tinnev-elly senna, 6 of godhaamla, 8 of silk cotton, 8 parts of virinchi satva, 5 parts of kanda, 3 parts of yellow orpimet, 7 parts of kaarmukha, or brown barked acacia?, these 26, powdered, puri-fied, and filled in crucible and placed in furnace and boiled to 800 degrees, and poured into Darpana yantra, will yield a fine roudrikaa-uarpana.

FOURTH CHAPTER

Shaktyadhikaranam: The Power.

Maharshi Bharadwaaja:

☞ **"Shaktayassapta" Sootra 1.**
"The power sources are seven."

Bodhaananda Vritti:

In this chapter the motive power of the vimaana is explained. In the functioning of the vimaana, there are 7 distinct operating forces. They are named udgamaa, panjaraa, sooryas-haktyapa-karshinee or that which extracts solar power, para-shaktyaakarshinee or that which extracts opposite forces, a set of 12 shaktis or forces, kuntinee, and moolashakti or primary force. At set spots in the vimaana, the motors which produce these 7 powers should be installed, duly wired and equipped with springs and wheels, as prescribed.

It is said in "Yantra-sarvasva:"

"The seven kinds of powers which are required for the Vimaana are produced by 7 motors which are named tundila, panjara, amshupa, apakarshaka, saandhaanika, daarpanika, and shaktiprasavaka. Each of these produces its specific power. Thus tundilaa produces udgamaa shakti, panjaraa produces the panjaraa shakti, shaktipaa produces the power which sucks solar power, apakarshaka produces the power which plucks the power of alien planes, sandhaana yantra produces the group of 12 forces, daarpanikaa produces kuntinee shakti, and shakti-prasava yantra produces the main motive power.

Shownaka-sootra also says:

"There are seven sources of power of the vimaana: fire, earth, air, sun, moon, water and sky. The seven kinds of powers are named udgamaa, panjaraa, solar heat absorber, alien force absorber, solar electric dozen, kuntinee, and primary force."

"Soudaaminee-kalaa" says:

Ma, la, ya, ra, sa, va, na constitute the seven vimanic forces. Ma is udgamaa, la is panjaraa, ya is solar heat absorber, ra is the solar dozen, sa is alien force absorber, va is kuntinee, and na is primary force.

Their actions are thus defined in "Kriyaa-saara":

"The ascent of the vimana is by udgamaa shakti. Its descent is by panjaraa-shakti. Solar heat absorbing is by shaktyapak-arshinee. Alien force restraining is by parashakty snatcher. Spectacular motion of the vimaana is by the vidyud-dwaadashaka-shakti. All these various activities are by the prime force of the vimana."

Vidyuddwaadashaka is thus explained in "Soudaaminee-kalaa":

"The spectacular motions of the vimaana are of 12 kinds. Their motive forces are also 12. The motions and the forces are, proceeding, shuddering, mounting, descending, circling, speeding, circumambulating, side-wise motion, receding, anti-clockwise motion, remaining motionless, and performing miscellaneous motions."

Maharshi Bharadwaaja:

☞ **"Shaktayah-pancha -iti-Narayanaha." Sootra 2.**
"Narayana holds that the forces are five only, and not twelve."

Bodhaananda Vritti:

Five forces are generated by the yantra or dynamo called Sadyojaata, and they produce all the spectacular motions of the vimana.

Says "Shakti sarvasva":

"The motions of a vimaana are five, Chaalana, Gaalana, Panjaraprerana, Vakraapasarpana, and Spectacular manoeuvring."

Maharshi Bharadwaaja:

☞ **"Chitrinyeveti sphotaayanah." Sootra 3.**
Sphotaayana holds that chitrinee is the sole shakti.

Bodhaananda Vritti:

Sphotaayana declares that the force called chitrinee shakti is the one which enables the vimana to perform spectacular manoeuvres.

"Shakti-sarvasva" says that both from experience and scientific knowledge Sphotaayana propounds the view that 32 various kinds of motions of the vimaana are solely by the power of Chitrinee-shakti.

"Kriyaa-saara " also states that Chitrinee force of the 17th quality is solely responsible for the 32 types of aeronautical motions.

Maharshi Bharadwaaja:

☞ **"Tadantarbhaaavaat Saptaiveti" Sootra 4.**
"The shaktis are 7 only, and include all others"

Bodhaananda Vritti:

Out of the five forces produced by the sadyojaata mechanism, panjaraa shakti is the most important. The other shaktis are incidental to it, just as sparks are incidental to fire. Chaalana and other motions may therefore be said to result from panjaraa shakti.

Says "Shaktibeeja": "It is by the panjaraa shakti generated by sadyojaata yantra that the chalana and other shaktis branch out.

"Shakti kousthubha" also says, "From the panjaraa shakti produced by sadyojaata, emanate the chaalama and other 4 shaktis."

Thus since the other shaktis branch out from panjaraa shakti, they may be said to be in essence identical with it. That panjaraa and chitrinee are included in the seven shaktis which have been enumerated by Maharshi Bharadwaaja. Hence there cannot be said to be any conflict of opinions. Some even hold the view that each one of the seven shaktis is capable of producing all the 32 motions of the vimaana. But since each of the several motions of the plane is definitely ascribed to a particular kind of force, it would be incorrect to hold that one force could be responsible for the whole gamut of motions. Any attempt to give practical effect to such a theory would prove disastrous. Therefore the right conclusion is that the seven forces are the true cause of the 32 kinds of aerial activities of the vimaana.

FIFTH CHAPTER

Yantraadhikaranam: Yantras: Machinery.

Maharshi Bharadwaaja:

☞ **"Athha Upayantraani." Sootra 1.**
"The Mechanical Contrivances."

Bodhaananda Vritti:

Having described the forces or energies required for the various functions of the vimaana, now the mechanisms necessary for these activities are described.

"Kriyaa-saara" says:

"As stated by the eminent Bharadwaaja in "Yantrasarvasva", the mechanical equipments necessary for the vimaana are 32. They are vishwakriyaadarsa or universal reflecting mirror, shaktyaakarshana yantra or force absorbing machine, pariveshakriyaayantra or halo- producing machine, angopasamhara yantra or machine for folding up or contracting its parts, vistrutakriyaa yantra, or expanding yantra, vyroopyadarpana or fantastic mirror, padmachakra-mukha, kuntinee shakti yantra and pushpinee shakti yantra, pinjula mirror, naalapanchaka and guhaa-garbhabhidha yantras, tamo-yantra or darkness spreading machine, pancha vaataskandhanaala, roudree mirror, vaataskandha naalakcelaka, vidyudyantra or electric generator, and shabdakendra mukha, vidyuddwaadashaka, praanakundalinee, shaktyudgama, vakraprasaarana, and shaktipanjara keelaka, shirah-keelaka and shabdaakarshana, pataprasaaranayantra, dishaampati yantra, pattikaabhraka yantra, suryashaktyapakarshana yantra or collector of solar energy, apasmaaradhooma prasaarana or ejector of poisonous fumes, stambhana yantra, and vyshwaanara naalayantra."

They are thus described in "yantrasarvasva," chapter 7, by the illustrious Maharshi Bharadwaaja.

Maharshi Bharadwaaja:

☞ "Athopayantraani." Sootra 1.
"Subsidiary Yantras."

Bodhaananda Vritti:

Prepare a square or circular base of 9 inches width with wood and glass, mark its centre, and from about an inch and half thereof draw lines to the edge in the 8 directions, fix 2 hinges in each of the lines in order to open and shut. In the centre erect a 6 inch pivot and four tubes, made of vishvodara metal, equipped with hinges and bands of iron, copper, brass or lead, and attach to the pegs in the lines in the several directions. The whole is to be covered.

Prepare a mirror of perfect finish and fix it to the danda or pivot. At the base of the pivot an electric yantra should be fixed. Crystal or glass beads should be fixed at the base, middle, and end of the pivot or by its side. The circular or goblet shaped mirror for attracting solar rays should be fixed at the foot of the pivot. To the west of it the image-reflector should be placed. Its operation is as follows:

First the pivot or pole should be stretched by moving the keelee or switch. The observation mirror should be fixed at its base. A vessel with mercury should be fixed at its bottom. In it a crystal bead with hole should be placed. Through the hole in the chemically purified bead, sensitive wires should be passed and attached to the end beads in various directions. At the middle of the pole, mustard cleaned solar mirror should be fixed. At the foot of the pole a vessel should be placed with liquid ruchaka salt. A crystal should be fixed in it with hinge and wiring. In the bottom centre should be placed a goblet-like circular mirror for attracting solar rays. To the west of it a reflecting mechanism should be placed. To the east of the liquid salt vessel, the elec-

tric generator should be placed and the wiring of the crystal attached to it. The current from both the yantras should be passed
to the crystal in the liquid ruchaka salt vessel. Eight parts of
sun-power in the solar reflector and 12 parts of electric power
should be passed through the crystal into the mercury and on
to the universal reflecting mirror. And then that mirror should
be focussed in the direction of the region which has to be pho-
tographed. The image which appears in the facing lens will then
be reflected through the crystal in the liquid salt solution. The
picture which will appear in the mirror will be true to life, and
enable the pilot to realise the conditions of the concerned region, and he can take appropriate action to ward off danger and
inflict damage on the enemy.

Next Shaktyaakarshana yantra:

"Yantra sarvasva" says, "Owing to the etherial waves and rag-
ing winds of the upper regions in accordance with die seasons,
evil forces are generated which tend to destroy the vimaana.
The Shaktyaakarshana yantra in the vimaana is meant to subdue
those forces and render them harmless."

Narayana also says:

"Three fierce forces arise from the fierce winds and ethereal
waves, and cruse destruction of the plane. The shalayaakarshana
yantra by its superior force subdues them and ensures safety of
the vimaana."

Its construction is as follows:

The base is to be 3 feet long and 2 feet wide, and made of
krouncha metal. A 12 inch tall 3 inch wide pole or peg made of
27th kind of glass should be fixed in its middle. To the east of
it, as also to the west, 3 centres should be marked on each side.
To the north and south also 2 centres should be marked on each
side. At each centre screw-bolts should be fixed. Then tubes
made of the 107th glass, with cleaned wiring should be fixed. A
goblet shaped 15 inch sized glass vessel should be fixed on the
base of the central peg. A 1 foot circular glass ball with three

holes should be fixed in the main centre. A triangular shaped 1 foot sized mirror made of Aadarsha glass should be fixed on the 3rd kendra. Two circular rods made of magnetic metal and copper should be fixed on the glass ball so as to cause friction when they revolve. To the west of it a globular ball made of vaatapaa glass with a wide open mouth should be fixed. Then a vessel made of shaktipaa glass, narrow at bottom, round in the middle, with narrow neck, and open mouth with 5 beaks should be fixed on the middle bolt. Similarly on the end bolt should be placed a vessel with sulphuric acid (bhraajaswad-draavaka). On the pegs on southern side 3 interlocked wheels should be fixed. On the north side liquefied mixture of load-stone, mercury, mica, and serpent-slough should be placed. And crystals should be placed at the requisite centres.

"Maniratnaakara" says that the shaktyaakarshana yantra should be equipped with 6 crystals known as Bhaaradwaaja, Sanjanika, Sourrya, Pingalaka, Shaktipanjaraka, and Pancha-jyotirgarbha.

The same work mentions where the crystals are to be located. The sourrya mani is to be placed in the vessel at the foot of the central pole, Sanjanika mani should be fixed at the middle of the triangular wall. Pingalaka mani is to be fixed in the wide mouthed glass globe. Bhaaradwaaja mani should be fixed in the opening in the naala-danda. Pancha-jyotirgarbha mani should be fixed in the sulphuric acid vessel, and Shakti-panjaraka mani should be placed in the mixture of magnet, mercury, mica, and serpent-slough. All the five crystals should be equipped with wires passing through glass tubes.

Wires should be passed from the centre in all directions. Then the triple wheels should be set in revolving motion, which will cause the two glass balls inside the glass case, to turn with increasing speed rubbing each other, the resulting friction generating a 100 degree power. That power should be conveyed through wires to the sanjanika mani. Mingling with the force existing therein, that force issues out and should be transmitted through wires to the sourrya mani. On contact of the power therein the force will split into 5 streams. Each of the five power streams should be connected with one of the manis, Bhaarad-waja, Sourrya, Pingala, Pancha-jyotirmani, and Shakti-panjara mani. Mingling with the force in each mani, they form five forces, which are named by Atri maharshi as Raja, Mourtvica, Chundeera, Shoonya, and Garbha-vishodara. These should be passed by wires to the sulphuric acid vessel. They then form 3 forces, named marthanda, rowhinee, and bhadra. Marthanda shakti should be passed into the load-stone, mercury, mica, and serpent slough liquids. The resulting current should then be passed through wires to the wide mouthed glass globular vessel. Solar force pregnant with etherial force should be passed into the Naaladanda, and thence to the vessel with marthanda shakti. The power of the solar rays entering that vessel mingles with the marthanda shakti inside, and the resultant force has to be fo-cussed towards the adverse force of the etherial current which will be thereby nullified and the vimaana will be protected.

Then the Rohinee shakti should be passed through wires into the vessel containing the fivefold load stone, mercury, mica, serpent slough acid, and the resulting current passed to the Bhrajasvaddraavaka or luminous acid vessel at the foot of the central pole. Then from the air- route collect the wind-force impregnated solar rays and pass them also into the above ves-sel. Mingling with the rowhinee shakti therein a super-force will be created which should be passed through the northern pivot, into the rowhinee power vessel. The united force should then he directed against the malefic wind force in the air-route, so that it will tame the evil force and protect the Vimaana.

Then from the suragha tube Bhadraa shakti should be passed into five fold acid vessel. The resulting force should be passed through wired tubes to the foot of the triangular wall, and thence to the pivot on the southern side. The force should then be directed against the evil roudree Force in the air-route. Neutralising that third destructive force in the sky, the vimaana will be allowed smooth passage in the sky.

The Parivesha-kriya yantra:

According to Yantra-sarvasva, by manipulating the five forces a halo is formed around the vimaana, and by drawing the solar rays into contact with it, the rays will speed the aeroplane along the rekhaamaarga or safety line. This is achieved by the operation of the above said yantra.

Narayanacharya also says:

"The mechanism which will manipulate the five forces so as to create a halo round the plane, and attracting the solar rays and contacting them with the plane, make them draw the plane smoothly and speedily along the air route without swerving into danger, is called parivesha-kriyaa yantra or halo-forming mechanism."

Soudaaminee kalaa says, "The forces of ksha, ja, la, bha, and ha, when united attract solar rays. "

According to "Gopatha-kaarika," the forces in shireesha or Indra or lightning, clouds, earth, stars; and sky, are indicated by the letters ksha, ja, la, bha, and ha. By combining those live forces a halo, like that around the solar orb, will be created, and it will have the power or attracting solar rays.

Kriyaa-saara says Shireesha has 2 parts, Clouds have 8 parts, Earth has 5, Stars have 7, and Aakaasha or Sky or Ether has 10. The Aakarshana yantra should attract these forces and unify them. Then through the mirror above the vimaana attract solar rays, and apply them to the unified forces,

A halo will be created, and that halo, in combination with the solar rays, will draw the plane through a safe course like a bird held by a string, Its formation is thus explained in Yantra-sarvasva:

"Athha Yantraangaani"

We now deal with the parts of the yantra:

A foot-plate: 23 main centres to be marked on it, with lines connecting the centres. Similar number of revolving screws, wired tubes, pole with three wheels, eight liquids, eight crystals, eight liquid containers, mirror to attract the forces of shireesha, cloud, earth, stars, and aakaasha, five electric mechanism, five barks of trees, copper coated wires, five leathers, hollow screws, revolving screw with wire, vessels for storing the energies, vessel for mixing the energies, smoke-spreading yantra, air-fanning yantra, halo-creating tube made of milky-leather, solar ray attracting mirror tube, tube for collecting the solar rays reflected in the mirror at the top portion of the vimaana, crest-crystal, screw for connecting the solar rays to the vimaana. These are the 23 parts of halo producing yantra.

Its construction is now explained: A wooden base 23 feet square, made of black pippala or holy fig tee. 23 centres enclosed in a case made of 35th type of glass. 23 lines to the centres. Revolving keys to be fixed at the 23 centres. Wired glass tubes should connect one centre with another. A glass pole made of the 37th type of glass, 5 feet long, 1 foot thick in the middle, 18 inches thick at the neck, with a 10 fact wide top, should be fixed as the central pillar, with 3 revolving wheels. Eight acids should be placed in the eight directions from the north-east side. Their names are rubnaka, kraantaja, taarkshya, naaga, gowree, vishandhaya, khadyota and jwalana.

The rubnaka acid is to be placed in the north-east centre, kraantaja in the centre, naaga at the southern centre, gowree at the south- west corner, vishandhaya in the western centre, khadyota at the north-west centre, and jwalana at the northern centre in 8 glass vessels.

The names of the vessels are also given by Shaarikaanaatha: shila, abhra, paara, vyrinchika, vaaluka, asuragranthika, sphutika, and pancha-mrith. The 8 vessels are made out of these 8 elements by process defined in "Darpana-prakarana."

The rubnaka acid should be filled in shila-darpana vessel; kaarshnya-acid in abhrakaadarsha; kraantaja acid in paaraadarsha vessel; naagadraava in vyrinchi-aadarsha vessel; khadyota acid in sphutikaadarsha; gowree acid should be filled in vaalukaadarsha vessel; vishandhaya acid should be filled in suragrathika vessel; and jwalana acid in panchamrid vessel.

In the 8 acid filled vessels 8 crystals are to be inserted. As mentioned in "Maniprakarana" their names are dhoomaasya, ghanagarbha, shalyaaka, shaarika, tushaasya, somaka, shankha, and amshupa.

Having mentioned their names, we now explain their disposal. Dhoomaasya mani is to be placed in rubna acid vessel. Ghanagarbha mani should be placed in kraantaja acid vessel. Shalyaaka in kaarshni acid vessel. Shaarika in naaga acid vessel. Tushaasya is to be placed in gowree acid, Shankha in jwalana acid; Somaka in vishandhaya acid; and Amshupa mani is to be placed in khadyota acid vessel.

In front of these manis, eight shaktyaakarshana, or energy-imbibing mirrors are to be fixed. Their names according to Bharadwaja are taaraasya, pavanaasya, dhoomaasya, vaarunaasya, jalagarbha, agnimitra, chaayaasya, and bhanukantaka. Their location is as follows: Six inches in front of dhoomasya mani the taaraasya mirror with an iron rod with a switch attached to it should be fixed. Pavanaasya mirror should be fixed similarly in front of ghanagarbha mani. Dhoomaasya mirror should be fixed 6 inches in front of shalyaaka mani. Vaarunaasya mirror should be fixed in front of shaarikaa mani. Jalagarbha mirror should be fixed in front of somaka mani. Agnimitra mirror should be fixed in front of tushaasya mani. Chhayaasya mirror should be fixed in front of shankha mani. And Bhanukantaka mirror should be fixed in front of amshupaa mani.

Then in the western centre should be installed the electric generator with switch. Copper-coated wires covered with live kinds of skins, should be spread all-round, proceeding from the shakti-yantra or electric generator. The names of the five skins, according to "Kriyaa- saara," are rhinoceros, tortoise, dog, rat or hare, and crocodile.

According to "Twangnirnaya-adhikaara," or chapter on skins, for seats in vimaanas, and, for containing acids, and covering wires, five kinds of skins are mentioned by the learned; skins of rhinoceros, tortoise, dog, rat or hare, and crocodile. These five are to be used for the purposes of cove-ring, and seating. Wires covered with these skins are good conductors of electricity. The bhraamanee keela, or central revolving pole should be fixed in the centre so that when it revolves all the other pivotal centres also revolve. Eight energy storing vessels should be placed in the 9th, 8th, 10th, 12th, 13th, 15th 16th and 11th centres. The sammelana vessel or coordinating vessel should be placed in the front of the 23rd centre. To the south of it at the 21st centre the wind blowing mechanism should be fixed.

The Vaata-prasaarana or wind-blowing yantra is thus described: In the central pivot there are to be 5 wheels which will turn with 100 linka revolutions by contact with electric wires: in the east and west two bellows on pivots: two air-containers with 3 mouths or openings: 6 wheels which prevent air-motion: two tubes with switches which will cause spreading: wheels with keys that will induce speed, or full speed, slow, very slow or stop, shaped like a tortoise, having two bharas or parts?, and having a wheel fixed at the top. That is a vaata-prasaarana yantra.

The dhooma-prasaarana yantra or smoke-spreading yantra is as follows: with three openings, 5 satchels inside, 8 wheels, three keelakas or switches, encircled by electric tube, provided with smoke-generating mani or crystal, and equipped with 5 acids, with two churning wheels with keys, with two smoke containers attached to the bellows tube, with smoke-spreading yantra, and it is to be fixed at the 20th centre.

The parivesha-kriyaa naala or halo-creating tube is thus explained. Out of 5 milks from 5 kinds of milk trees, 6 barks of trees, and 2 valkalas (hemp, jute), cloth is fashioned. And that cloth should be used in preparing the parivesha-kriya or halo-forming tube.

It is stated in "Ksheeree-pata kalpa":

In the realm of milk-yielding trees, dugdha-pranaalee, pata-paadapa, payodharee, panchavatee, and virinchi are the 5 most suitable for manufacture of milk-cloth useful for vimaanas.

"Patapradeepikaa" also says, "Among the milk-trees, the best for producing milk-cloth are the following five, payodharee, panchavatee, viranchi, patapaadapa and dugdhapranaalika.

The six bark-trees are godaakanda, kurangaka-niryaasa, aan-dolikaaviyatsaara, lavika, prishatka, and kshmaamala. In conjunction with the milk from milk-trees these barks produce cloth which is flawless, strong, and soft.

For the two valkalas, according to "Agatatva-nirnaya" out of 5000 kinds of valkalas from shaarikaa to panchamukhee, the two named simhikaa and panchaanga are said to be excellent for producing the milk-cloth required for vimaanas.

The composition of the cloth is as follows:

Dudgdhapranaalika milk 8 parts, 10 parts of the milk juice of the patavriksha, 7 parts of payodaree or cocoanut milk, 18 parts of the milk of the 5 vata or ficus trees, and 12 parts of virancha tree.

The ambikaa-shatka composition is 10 parts of godaa-kan-da, 17 parts of gum from kurangaka, 15 parts of aandolikaa viyatsaara, 12 parts lavika, 20 parts of prishatka, and 15 parts of kshmaamala.

The two jute cloth proportions are given in "Shana-nirnaya chandrikaa," as 28 parts of simhikaa jute, and 18 parts of pan-chaangavalkala jute.

These proportions of 5 ksheera or milk, 6 ambika or barks, and 2 valkalas or jutes, should be mixed together and unified, and boiled in paakaadhaana yantra and churned a number of times, and processing with acids 12 times, should be filled in pata-garbha kriya or cloth-making yantra, and milk-cloth of excellent quality obtained. The parivesha kriyaa-tube made out of this cloth will, by manipulation of the concerned switch, expel smoke from the vimaana, and by quick advancing and reverse revolutions of the wheel will spread the smoke all round so as to envelope the vimaana by means of the smoke-screen.

The Kiranaakarsha-Naala:

16 parts of the 305th variety of glass, 5 parts of kaancho-likaabharana, 6 parts of nagakesara or merua ferrea,—aletris hy-acinthoides, 4 parts of couries, sunflower, and Indian spikenard, 8 parts of pure borax, iron dross, onion juice, cuscus grass pow-der, ruby glass, the three varieties of salt-petre, sand, essence of suranjikaa, viranchi flour, essence of black-mica, essence of bael fruit, and juice of flower buds, these twelve ingredients, in the proportion of 27, 5, 7, 3, 8, 7, 3, 11, 8, and 12, are to be filled in the frog-shaped crucible, and placed in the frog-shaped furnace, and melted with 300 degrees of heat with the help of two-winged bellows. The resulting liquid is to be poured into the darpana yantra or glass-making machine, so as to produce the kiranaakarshana or rays-attracting yantra.

The tube made of this glass should be fixed at the top of the concerned yantra.

Next- the pratibimba-arka-kiranaakarshana naala, or tube for attracting the reflection of the solar says:

According to "Naalikaa-nirnaya," the essence of squash gourd, juice of momardica, 2 parts, of the salt of the two wheeled root vegetable, 3 parts of salt of simhamoola, 122nd type of glass, essence of white mica, jelly stone, borax, root of Bengal-madder, thorn at the root of bamboo, lead, mercury, these 15 ingredients are to be mixed in the proportion of 5, 12, 4, 3, 7, 3, 11, 4, 9, 12, 20, 18, 12, 5, 20. The mixture should be filled in the crucible known as samavargika, and heated in the furnace of the same name, and heated to the degree of 315, with the aid of bellows called suraghaa. The resulting liquid should be poured into the mirror—making machine. The resulting product will be a fine bimbaarka-kiranaadarsha, or reflected solar ray attracting mirror. This should be fixed in the central portion of the vimaana and in the 10th kendra, with five circled screws.

Now we deal with the crest crystal of the vimaana. The crest-crystals are of 103 kinds. They are named in "Mani-kal-pa-pradeepika" as belonging to the 12th class of 32 groups of crystals. Their names are shankara, shaantaka, kharva, bhaas-kara, Mandana, kalaantaka, deeptaka, nandaka, chakrakantha, panchanetra, Rajamukha, Raakaasya, kaalabhyrava, chinta-mani, koushika, chitraka, bhaskara, uduraaja, viraaja, kalpaka, kaamikodbhava, panchasheershna, paarvanika, panchaaksha, paaribhadraka, isheeka, kaashabhrit, kaala, kanjaasya, kowtika, kalaakara, kaarmika, vishaghna, panchapaavaka, symhikeya, rou-dramukha, manjeera, dimbhika, pingala, karnika, krodha, kra-vyaada, kaala-kowlika, vinaayaka, vishwamukha, paavakaasya, kapaalaka, vijaya, viplava, praanajanghika, kaarmukha, prithu, shinjeera, shibika, chanda, jambaala, kutilormika, jrimbhaka, shaakamitra, vishalya, kanka-gowrabha, suragha, suryamitra, shashaka, shaakala, shaktyaakara, shaambhavika, shibika, shuka, bherunda, mundaka, kaarshnya, puruhoota, puranjaya, jam-baalika, sharngika, jambeera, ghanavarshmaka, chanchvaaka, chaapaka, ananga, pishanga, vaarshika. Raajaraaja, naagamukha, sudhaakara, vibhakara, trinetra, bhoorjaka, kumuda, koorma, kaarmuka, kapila, granthika, paashadhara, damaruga, ravi, mun-jaka, bhadraka.

These are the 103 crystals suitable for being fixed as crest-jewels of the vimaana. One of them is to be fitted to the cen-tral pinnacle at the top of the vimaana, and the wires from the electric dynamo should be connected to it, so that it might be supplied with power. On the upper side should be attached wires for collecting solar rays, so that the two forces might act in combination.

The switch-gear for connecting the vimaana with the solar energy is explained in "Brihath-kaandika." Sandhaana-keelakaas are of 25 kinds. Their names are pinjuleeka, keeranaka, dimb-

haka, paarvateeyaka, kachchapa, gaaruda, uddanda, shak-
tipa, govidaaraka, pavanaasya, panchavaktra, vajraka, kankana,
ahirbudhnya, kundalika, naakula, oornanaabhika, trimukha,
saptasheershanya, panchaavartha, paraavatha, aavarta, naabhika,
oordhvaasya, shilaavarta.

Amongst these the 9th, govidaaraka, is best suited to connect the vimaana with the solar beams for safe navigation. This
is Pariveshakriyaayantra.

Next Angopasamhaara yantra:

During the passage of sun and other planets in the 12 hous-
es of the zodiac, owing to the varying speeds of their progres-
sive and retrogressive motions, conflicting forces are generated
in the zodiacal regions, and their collisions will let loose floods
of fierce forces which will reduce to ashes the parts of the plane
which get involved with them. The pilot should get warned by
the ushna-pramaapaka yantra, or heat-measuring instrument,
and quickly fold the concerned parts and ensure their safety.

It is described in "Yantrasarvasva" as follows:

Purifying the metal sumrileeka mixed with manjeera, a pedestal should be cast, 12 feet long, 18 inches thick, and shaped as
a square or circle. Then mixing the magnetic stone and dimbika,
after purifying them with acids, cast a pole 3 feet thick and 30
feet tall, with springs, as in an umbrella, at the foot, in the middle
and at the upper end, and fix it in the centre of the pedestal.
Rods made of mixed metal like umbrella rods, provided with 5
springs, should connect the springs in the pole with the several
limb mechanisms of the vimaana. Two revolving wheel springs
with two tubes with 3 faces and 3 wheeled springs should be
fixed at the bottom of the pole, near the spring. Above there
should be fixed an oiling tube which will keep all the springs
well-oiled. When a particular limb of the plane has to be contracted the spring at the foot of the pole should be turned so as
to induce the spring of the part to operate so as to contract or
open up the part as need be so that the danger to the part will be

prevented. By the operation of the angopasamhaara yantra, any part of the plane can be folded up to avoid danger and opened out subsequently.

Vistritaasyaa kriyaa yantra or wide-opening mechanism:

When the various powers, subterranean, eight cardinal points, earth, cloud, electricity, and oceanic, consemble in pad-ma-mukha, a power called vishambharee is generated. It breaks through the earth, emitting great heat, mounts with a 300 linka speed to the upper sky regions, and reaching the aerial routes, envelopes the vimaana, and affects the personnel inside causing grave physical disabilities, and paralysing the brain. For the purpose of curbing it and nullifying it, the vistritaasyakriyaa yantra is to be installed in the vimaana.

According to "Yantrasarvasva," a foot-plate, of an arm's length, and 22 inches thick, and round-shaped, is to be made of the wood of the sacred peepul tree. A pole of an arm's thickness, and 32 inches high, is to be fixed in the middle of it. Reversible wheeled double- switches should be fixed along its height, connecting each of the sectional mechanisms in the vimaana, through tubes reaching to the bhastrikaa naala or bellows tube attached to the mechanisms. At the foot of the pole three revolving wheels, and at its back the contracting switches, have to be fixed.

First peetha or footplate, then pillar, then revolving springs, jointure tubes, two-wheeled keelakas, two-winged bellows, three wheel moving mechanism, contracting mechanism, are eight constituent parts of this machine.

First the triple wheeled mechanism should be switched on. That will set the double wheels in motion. That will make all the springs attached to the pillar begin to operate. The two winged bellows attached to the double-wheels will open up. Wind will rush out and force through all the sandhi-naalas or jointure tubes. That will set the bellows in the central operating; thereby the bellows of the sectional mechanisms will come into play, and air will flow out in a flood, and taking hold of the vishamb-haraa shakti expel it to the aerial regions where it will get lost. Thus the personnel inside the vimaana will be saved from dis-abilities and restored to normalcy.

Vyroopya mirror: Says "Yantrasarvasva",

When enemy planes come intent on destroying the vimaana, the vyroopya mirror is intended to frustrate them. Its parts are, peetha or stand, central switch-gear, electric pole, smoke tube, betel-nut oil, triple-wheeled spring, three satchels, smoke light, and contraction tube.

The peetha or seat should be 2 feet wide and 2 feet tall, and circular, and made of bael tree wood. 12 centres are to be marked therein. At each centre revolving joints should be fixed. Jyotistambha or electric pole, 24 inches thick and 24 inches tall and made of vyroopya darpana glass, is to be fixed in the centre. In front of it the electric machine should be fixed in the 2nd kendra. In the 3rd kendra should be fixed the turning smoke tubes with winding wires. The oil vessel should be fixed in the 5th kendra. The 3 satchels, with 3 mouths, one foot high and made of milk-leather should be fixed in the 6th 7th, 8th and 9th kendras, up to the smoke tube. In the tenth kendra should be fixed the smoke-extinguishing tube mechanism, and the light-extinguishing tube in the eleventh kendra. The winding wire tube should be fixed in the 12th kendra.

The operation of the mechanism is as follows:

Drawing the electrical energy from the dynamo, it must be applied to the triple-wheeled mechanism. That will be set in mo-tion. The wires proceeding from there will convey the power to all the other mechanisms and set them in motion. Kendras 3, 4, and 5, will become active. When kendra, 9 is switched on the koshas attached thereto will become active. From the 5th kendra the current should be passed to the oil vessel. The oil will then convert itself into poisonous .gas. The gas should be filled in the 3 satchels and the 3 tubes. The fumes from two of the tubes should then be discharged towards the enemy planes. They will encircle the enemy planes and envelope them with a smoke-screen. Then the betelnut oil should be lighted, and fluxed in the jyoti stamhha or light-pillar. The light within the pillar will suffuse it with red glow like a china rose; and pervade the sky. Then the electric glow should be applied to that glow. The resulting glow will be multi-coloured like a rain-bow, with violet, indigo, blue, green, yellow, orange, and red. Then the

poison-fumes from the 3rd tube should be drawn through the air tube, and let into the multicolour-glowing light-pillar. The fume will burst into light, and then should be passed through tube into the vyroopya-darpana. The light glow will pervade the mirror and attain 3000 degree intensity, causing a blinding glare and paralising the enemy. Then the gas in the three satchels should be projected with 25 linka speed towards the smoke screen enveloping the enemy. Then the smoke from the tubes should be projected with 28 linka speed into that screen. Then the smoke filled glow will flood over the enemy personnel and affect their body joints, organs, mind, vision, and induce inertness, and make them all fall down senseless. Then the pilot could change his air-route and proceed forward safely.

Then Padmachakra mukha yantra:

According to "Yantra sarvasva," its parts are, peetha or pedestal, pillar, tubular pole, electric wiring, glass lotus petals, lotus formation process, places where the lotuses are to be located, wind inhaling and leather-bellows mechanism, contracting and expanding switches, triple-wheel fixing arrangement, air flow outlets, folding up mechanism. These are the 12 parts of the padmachakra mukha yantra.

The peetha or seat should be made of the wood of pippala or the holy fig tree, 8 feet and 3 feet high, and square or circular. Mark 12 fixing centres on it. From the central pillar draw lines towards the 12 spots. The central pole, two tubular posts on either side of it, electric wire in eastern centre, lotus petals in the north, formation of lotus in the northern and southern centres, fixing of the lotuses from the north-east to the south- east corner, to the east air-filling bellows. In the north west corner the contracting switch, and the expansion switch in the south-west corner, triple-wheel revolving mechanism on the eastern side, air flow outlets underneath each lotus. To its south, the contraction switch.

These are the 12 parts to be fixed in the 12 centres.

The production of the parts is as follows:

The central pivot should be made of abhra-mrid-darpana, or mica-sand glass. According to "Darpana-prakarana"—

5 parts of rambhasatva (plantain stem?), 8 of manjoosha (madder root?), 5 parts of kaanta (ayaskaanta? sooryakaanta?), 8 parts of kravyaada (jataamaamsi), 3 parts of aadhaka essence, 7 parts of tortoise shell essence, 18 of bhalyatvak, 3 essence of kudmala or flower buds, 8 of bamboo salt, 3 of hooves, 28 parts of shoonya-mrid or mica ash, 4 of trivikrama kshaara, 2 of conch, 5 of mercury, 8 of salts, 1 of creepers, 3 of silver, 3 of eye-oint-ment, these 18 ingredients, purified, filled in crucible, placed in varaatakunda furnace and boiled to 200 degrees, and slowly filled in darpana yantra, will yield an excellent abhra-mrid darpana.

Two tubular poles of the size of an arm, made of this glass, should be placed on either side of the pivot. From the central pole electrical wiring should be connected to the 12 centres. In the centre of the switch tubes should be placed the lotus petals, and 150 finely made glass lotus petals should be spread on the northern side electric wires.

The petals are to be made, according to Lalla, by mixing 15 parts of the mica glass, with 4 parts of sourika salt, duly mixed and finely powdered and melted in pattikaa machine, when like onion-skin layers, petals will take shape. Then the wires attached to the petals should be brought together from the several cen-tres, and attached to the lotus forming mechanism. By turning the concerned wheel the petals will move towards the centre and form a lotus. Each petal will then become a tube, and by their juggling each tube will form 2 petals. The air-attracting mechanism should be placed in front and set to work. With a shrieking noise the air will be sucked in by each tube and the petals will shoot the air far into the outer air.

It is said in "sandhaana-patala—"

The scattering of a blizzard which may obstruct the progress of the vimaana is only possible by means of the padma-sand-haana and not otherwise. Therefore the spots where the lotuses are to be inserted are now indicated. On the eastern side from the north east to the south east the lotuses are to be erected in seven places in close order. Beneath the seven lotuses should be fixed seven leathern bellows capable of deep draughts of air. On the north west corner should be fixed the double-wheeled contracting mechanism.

According to "Kriyaa saara," by turning the main wheel in right motion, and the upper wheel in reverse motion at full speed, the yantra will suffer contraction. This machine is composed of 6 wheels spread out, 5 naalaas or tubes, 12 wires and 12 openings, and 12 keys which will cause contraction of the 12 parts, with widened mouth at the upper and lower parts, and provided with 2 revolving keys. By placing such a contracting machine in the north-west corner, the machine could be contracted when desired.

Now we shall deal with the expanding mechanism. It is round like a water pot, with 12 wheels and mouths, having 12 tubes with rods inside with 12 revolving springs for ascending motion, and with a central spring for filling with air. With such a mechanism the yantra can be made to stretch its parts. This should be fixed in the south west corner.

Then at the eastern face the triple-wheeled revolving spring, called "bhraamanee-keelaka", should be fixed.

It has 3 ivory wheels, consists of 3 poles, wooden top shaped like shimshumaara, with wheels with spring on top. By its operation the several parts of the yantra are set in motion, and by the operation of the concerned springs, the yantra will expand. Therefore the 3 wheeled bhraamanee mechanism should be properly fixed at the eastern kendra with 5 bolts.

Underneath the lotuses air flow routes should be provided. There should be openings 12 inches wide, 2 inches high, be leather- covered, made of pippala wood, with 7 tubes for the flow of wind. Seven such tubes should be fixed beneath the seven lotuses, and provided with keys.

In the southern centre the contracting mechanism or upas-amhaara keela, with 12 outlets, should be fixed.

Owing to the seasonal changes forces will generate in the joints of the outer space, and combining with the oceanic forces will reach the realm of air and cause a commotion which will spread out with fierce force into the farthest air pockets, and let loose typhoons which reach the vimaana, and produce a dusty excrescence which will induce chicken-pox-like skin eruptions on the pilots and other occupants, and also break up the vi-maana. In order to suck up that foul wind-flow, and expel it out of the vimaana, the padma-patra- mukha yantra is prescribed.

Next Kuntinee-shakti-yantra:

Now we shall deal with kuntinee-shakti yantra. In mid-sum-mer, out of the myriad heat rays of the sun, by the union of the 3, 5, and 10th class of rays, a fierce force of blazing heat named kulakaa is generated.

It is said in "Ritukalpa",

From the solar heat generative source 3 Mahaakshoni and 21 crores 500 lakhs 16 thousand and nineteen heat rays emanate. They are classified into 5 crores 8 thousand and 107 groups in Vaalmeeki ganita. Each group is divisible into 100 sub-groups. Of these when the rays of sub-groups 3, 5, 10 from the second group get mixed up in the heart of summer, a force called Kulakaa with fiery intensity is generated; and when it moves into the path of the flying vimaana, the plane will be reduced to ashes. To protect against that the kuntinee-shakti yantra should be installed in the neck portion of the vimaana.

Sage Narayana also says:

Amongst the divisions of the heat rays of the summer sun, the second group has 85000 rays. Out of them those numbered 8, 3, and 10 are specially intense, and they attract the pramlochana shakti from koorma portion of the universe, and produce a fierce heat-wave called kulikaa. If a vimaana happens to encounter it in its course, it will be burnt to ashes. To safeguard against that the kuntinee-shakti yantra should be installed in the neck section of the vimaana.

Lallaachaarya also confirms:

Out of the many groups of the heat-rays of summer, numbers 3, 5, and 10 in the 32nd division of the 2nd group of rays' tend to contact the pramlochana shakti in koorma and produce a fierce force called kulakaa which will destroy the vimaana. The erection of the kuntinee-shakti yantra in the vimaana will prevent it from such destruction.

According to "Yantra Sarvasva",

Among the constituent yantras of the vimaana, the kuntinee-shakti yantra is required to protect it from the combustible heat waves known as kulikaa in summer. Its parts are ground-plate, central switch-board, acid vessel cloth, with folds, chakra-danti naala, milk cloth, tube covering switches, revolving wheel equipped with electric wiring, and contracting mechanism.

The peetha or ground-plate should be 3 feet wide and ½ foot high, and round like a drinking bowl, seven kendras or centres commencing from the eastern side, turning switches in the seven centres, the acid vessel in the central kendra.

"Kriyaasaara" Says:

For capturing kulikaa the oil from gunja or the seeds of the shrub abrus and tobacco leaves, and mercury and shanaka crystal are recommended for use. The oils or acids of the seeds and tobacco leaves are to be filled in goblet like cup made of glass made of naaga, crownchika, and sowrambha metal, add purified mercury, and fix in the central kendra. Then apply the solar rays to the vessel. By the action of the rays on the acids the crystal in the vessel will become charged with a cold frigid force called krownchinee. Then when the kulikaa force enters the vessel with its fierce heat, it will be sucked in by the cold-storage crystal.

In the left kendra the cloth with folds should be fixed. Says "Patakalpa.—"

In order to confine in the crystal the fierce heat of kulikaa it should be wrapped in the folds of a cloth of fine and strong texture made of spikenard and jute yarn, with 5 folds and 3 openings. From the openings 3 glass tubes should be projected with downward bends into 3 wide mouthed vessels. To the north—east must be fixed the chakradanti naala for attracting the kulikaa force. Snake-skin, gum of srini, woollen yarn, soft grass, should be boiled together and lac-coloured cloth-like glass prepared, and purified with sundikaa wood oil. It should he rolled in coils just as a snake circles up in coils and sleeps. The tiny glass tubes should be attached at the bottom of the chakradanti as directed.

Then ksheeree-pata naala, or milk-cloth tube is to be fixed. Made of milk-cloth with wide-opening, strong, soft, a tube should he inserted in the mouth of the chakradanti, and its end should be made to reach the hole in the peetha. Through that the kulikaa force makes its exit. After placing ksheeree pata naala like this with key, the electric wire connected central operating switch should be placed in the west. And to the north-east of it the vistritaasya or opening out switch should be fixed.

Says "Kriyaasaara":

It should have two satchels, two openings, two right-revolving and reverse-revolving wheels. In the eastern opening should be fixed the 2 right-revolving wheels. And in the northern mouth should be fixed the 2 reverse motion wheels. And as in an umbrella, sticks connecting all the parts with the centre, for the purpose of expansion and contraction by turning a switch. By operating the switch in the eastern opening all parts will open out or expand. By operation of the northern switch all parts will close up. This is the upasamhaara keelakaa.

Having enumerated the parts of the yantra, their operation is now given. First the electric switch. By putting it on, the Bhraamanee chakra or pivotal wheel will revolve setting in motion individual parts as and when desired by turning their respective switches. Then electric current should be passed to the acid containing the crystal. Solar rays also should then be passed into it. Thereby, in the acid there will be generated a female shakti of 5 nyankas called sowlikaa. Similarly in the crystal there will be generated a male shakti of 8 nyankas called chulikaa. By operation of the electric current the two shaktis will get unified and produce an extremely cold shakti called "crownchinee," capable of attracting the kulikaa. That crownchinee force should be projected through naala or tube towards the kulikaa, like imbedding a gunja pea in a lump of lac. Thereby the crownchinee will drag the kulikaa inside the yantra through the tube and drop it into the acid vessel where it will be imbibed by the crystal.

Then the patormikaa key should be turned, whereby the patormikaa will become wide open preventing any air from entering the crystal by covering it completely. Then the chakradanti key should be turned slowly, so that its mouth opens out and sucks the hot kulikaa from the crystal, and stores it inside itself. Then the key of the sookshmaadarsa naala, fine mirror tubes, should be operated. The kulikaa in chakradanti will emerge through the 3 tubes. Then the vistritaasya key should be operated quickly so that all the parts will open out, and the kulikaa shakti will get out and disappear, and the danger to the pilot will have passed. Then by operating the upasamhaara keelaka, the expanded parts will close up and the yantra will return to normalcy.

Now we shall deal with Pushpinee yantra. When the pilot has to travel during spring and summer months, the pushpinee yantra is intended to provide him with necessary comforts.

According to "Khetavilaasa":

In spring a force called sowrikaa emanates from the southeast. And in summer a force called panchashikhaa arises in the north-west and is intensified by the sun's rays. Panchashikhaa contains two kinds of poisons. Sowrikaa having fire and moon contents is cold and hot, cold internally, and hot externally. It generates warmth in all creations, making the human kind perspire, and the trees and vegetation bring out their milk and gums. Thereby their bodies are relieved of harmful materials likely to lead to diseases.

By its cold effect and attracting the spring effect from the solar rays, it permeates all things, and brings out shoots, tendrils, flowers and a glow in all trees and creepers. Similarly it effects the 7 physiological components of the human body and increases their vigour, strength, growth, and glow.

Panchashikhaa shakti or force effects movable and immovable life adversely by its stultifying influence, shrinks and dries up the growth process of both vegetable and animal life and causes deterioration. To counteract this harmful effect of the season on the personnel of the vimaana, the pushpinee yantra is commended as one of the constituents of the aeroplane.

Its parts are, the base, the cold processing mirror, keelaka or key, cold generating crystal, acid vessel, electric wheel with 100 spokes.

The sunda-mud made glass is prepared, according to "Paarthiva-paaka Kalpa" as follows. Take salt, shinjeera, bone, and betel-nut salt, durona, kuruvinda grass (cyperus rotundus), gum, sowraashtra mud, virinchi vatika or banyan bark, silk cotton tree bark, and coir salt, these ingredients are to be taken in the proportion of 5, 12, 2, 3, 8, 3, 30, and 6, purified, filled in the crucible, and placed in the tortoise shaped furnace, boiled 32 times in 100 degree heat with the help of two faced bellows, and the resulting fluid poured into the cooling yantra. A pure and fine sunda-mud-glass will be formed.

With the glass thus produced by boiling 32 times, a base is to be formed 12 inches wide, 3 inches high, four-square or circular. From the centre of it 4 kendras or centres are to be marked. In the centre an arm-sized pivot made of the said glass should be fixed. On top of it is to be fixed the cold-processing mirror key. At its centre should be fixed the cold producing crystal. At the eastern centre should be placed the acid vessel.

Dravapaatra or acid vessel is described in "Kriyaasaara." It should be 12 inches wide and 12 inches high, shaped like a tumbler, circular, and hard like a cocoanut shell, and be made of the sheeta-ranjikaa glass.

The glass is described in "Darpana Prakarana". Shasha-piththa, udupiththa, borax, kutmala, jyotsnaa saara, rasonta kanda flour, kudupa-salt, mica salt, shoundeera jangha shalya flour, vaatohara, white niryaasa earth salt, and uragha.

These 12 ingredients should be taken in the proportion of 5, 3, 5, 1, 10, 10, 11, 8, 7, 2, 20, and 6, and after properly purifying them, fill them in lotus-shaped crucible, and placing it in the lotus shaped furnace filled with burning charcoal, and with the aid of the five-mouthed bellows blow the heat to 323 degrees temperature, and pour the liquid into the yantra. The resulting glass is called sheeta-ranjikaadarsa or cold-receptacle glass.

Cold-producing crystal is described in "Maniprakarana": 5 parts of couries and manjula powder, 4 parts udumbara salt, 3 of rubhna, 8 parts of varchulaka, 7 of sheeta ranjikaadarsha, 3 of vatu, 28 of shaalmali, 3 of salts, 7 of mercury, 8 of white mica, 8 of karkataanghri salt, 5 of chowlika satva, 15 of niryaasa earth, 25 of sampaathi bird kneebone—

These 14 ingredients, in the named proportions are to be purified and filled in mritkundala-moosha or earthen crucible,

and placed in kulakundika furnace, and with the aid of tryambaka bellows blown into 300 degree temperature. Fill the boiled
liquid into the mani- prasoothika, or crystal forming yantra. The
crystal produced will be pure, hard, and intensely cold.

In front of it should be fixed .the electric panka wheel, with
100 spokes and electric wiring, and purified by 3 acids.

As per "Kriyaasaara," 12 parts of copper, 3 of collyrium, 8
of zinc, should be mixed and melted with 100 degree heat. It
will become pure like gold, yellow, fine, soft, and strong. It is
called pancha-loha or five-in-one metal by those who know. 100
leaves like those of lotus should be formed out of them. Then
3 navels, three navel keys, and 3 wires, and a sounding keelaka or
key, or switch, or wheel.

First the navel wheels with hinged rods should be fixed so
that the 100 petals will he made to revolve with due speed on
the four sides of the wheel. Similarly by the side of the wheel
in front of it, another 100 petals should be properly fixed for
revolving in reverse direction. And electric wires should be fixed
on both sides of the centre of the western wheel, for operating
the 100 spoked electric panka or fan. Then the vessel should
be filled with the cold generating acid. And encircling the cold-
generating mani or crystal, it should be placed in another vessel
in the centre. And copper wiring enclosed in milk-cloth should
be attached to the wire in the acid vessel. Two wires from there
with switches should be connected with the cold ranjikaa glass
or mirror in right-turning fashion. Then current should be
switched into the electric wiring in the crystal and acid. Then by
the contact of the electric current the forces within the crystal
and the acid will get active and their combined cooling and com-
forting quality will enter the cold ranjanikaa mirror and con-
centrate in it. On operating the switch attached to it, the cold
effect will spread out all over the interior of the vimaana, and
overcoming the scorching seasonal effect, make it comfortable
and pleasant for the pilot, and restore his efficiency. Similarly the
100 spoked panka (fan?) should be switched on, when a breeze
will be generated and air-condition the atmosphere of the pilots.

Thus by the use of the crystal, acid, and panka, a state of pleasant comfort will be induced, and vigour, exhilaration, and competence will be injected into all the limbs of the body. Therefore this Pushpinee yantra should be installed in the southern section of the vimaana.

Next Pinjula Aadarsha or Pinjulaa Mirror:

By the collision of two winds giving rise to a whirlwind, and the fierce solar ray dashing against it, a lightning bolt erupts and strikes the unwary vimaana. To protect against such an event, the pinjula mirror is to be installed. An eight petalled lotus is to be made of the pinjula glass. Where the petals join, a circular dandaakaara should be made. At the back two hinged bolts should be fixed. They should be wound round by wires from the cold mirror. The back should be covered with a coir-made cloth covering. It should be fixed in the southern side of the vimaana, at an arm's height, facing the sun. The lightning will be absorbed by the projecting rods coiled with wires from the cold aadarsha mirror, and no evil effect will occur, and the pilot can proceed in safety.

And Naalapanchaka or Five tubes:

If the smoke from the kitchen over of the vimaana spreads, it will cause discomfort for people inside. Therefore the five tubes or pipes should be inserted for the smoke to go out and the air become clear. The pipes are to be manufactured as follows. Magnetic iron, pinjula mica, ghontaara metal, dhoomapaasya metal, and tortoise shell, are to be taken in the proportion of 1, 7, 5, 5, 8, purified, filled in crucible, and melted with 100 degree heat, and when ultimately cooled, a fine metal called vaataayanee metal, or window metal will result shining like gold.

With that metal 5 tubular outlets, 12 inches in diameter and 12 inches in length, should be fashioned. At one end of each of the tubes should be fixed a smoke-absorbing crystal. The tubes should be inserted in the 4 sides of the vimaana, forming outlets. One tube should be fixed at the ceiling. The dhoomapa crystals will attract the smoke and pass it to the outside, and clear the vimaana of its discomfort. Hence the necessity for the naalapanchaka, or five tubes.

Then Guhaa-garbha aadarsha yantra, or hidden mine discovering mechanism:
According to "Yantra Sarvasva" enemies would have placed mines and bombs underground for the destruction of the vimaana, unless they are discovered and de-fused in time there would be danger. Therefore the mine-finder yantra has to be installed in the vimaana.

Says kriyaasaara, out of the 72nd type of glass, make a triangular, a circular, and a quadrangular shaped glass mirrors. These are to be fixed as follows with bolts made of pancha-dhaaraa metal in a frame made of the wood of the anjishtha tree. The circular mirror should be fixed at the bottom facing downwards. The quadrangular mirror should be fixed facing upwards. The triangular mirror should be fixed to the west of these two, with a panchamukha keelee or 5 faced hinge. From the main pivot of the quadrangular mirror to the foot of the bolt at the south-east corner of the yantra, wires made of copper, tiles, and panchaasya metal should be drawn and connect them, and then the wire ends and chumbaka crystal should be placed in the mercuric-sulphur acid vessel. Four other wires should be made to circle the triangular mirror, pass through the mirror facing upwards, and fixed to the centre of the down-ward facing mirror. Then solar rays should be let in from the western side. A screen cloth coated with mirror-like gum should be placed opposite to the triangular mirror. Then the solar rays and electric current should be passed into the acid vessel containing the crystal. When the electrified rays from the crystal are passed on to the downward facing mirror, they will explore the ground over which the vimaana is to pass, and discover mines and bombs like mahagola and agni-garbha, which may have been inserted there and reflect their complete picture in the crystal in the acid vessel. The picture will then be projected to the screen opposite in clear detail, and by washing with chemicals present a perfect photograph of the buried mines and bombs, which could then be destroyed by due safety measures. Therefore the guhaa-garbha aadarsha yantra or mine-discovering yantra is essential for a vimaana.

Its parts are as follows:

First the 72nd type of mirror, known as suranjitaadarsha. "Darpana Prakarana" says:

Madder-root, live coal, ox-gall, snake-gourd, mercury, karanja or galedupia arborea, copper, 3 kinds of sharkara (sugar or sand?), borax, sulphur, chaaru or silk-cotton bark, lac, kuranga, rouhinee, iron-rust, panchaanana, liquid amber, Shiva or brionia laciniosa, vishwa, mica, paarvanija, vydoorya gem stone, in the proportion of 11, 27, 5, 7, 7, 3, 7, 5, 20, 3, 7, 3, 1, 32, 30, 38, 8, 7, 3, 9, 30, duly pulverised and filling in a beaked crucible, placed in a vaaraaha furnace, and heated to the 100th degree with the aid of the tortoise- shaped bellows. When the finely boiled liquid is poured in the cooling yantra, suranjika glass of exquisite quality will result, out of which three mirrors have to be fashioned for the guhaa-garbha-aadarsha yantra.

Aanjishtha Tree

Kriyaasaara says, "Many kinds of trees are suitable for use in making yantras. Of them all the tree called aanjishtha is the finest." "The trees having 5 qualities are 87 in number. The best among them all is aanjishtha," says "Udbhijya tatva saaraayanee."

Agatatvalaharee also says, the five qualities such as the capacity to capture reflections, and others, are found inherent in the
Aanjishtha (or madder root) tree. Therefore out of all woods the wood of that tree is most suitable for use in this yantra.

Pancha-dhaara-loha

In making yantras, pivots of various metals are being used. But for use in connection with the guhaa-garbha-aadarsha, or hidden mine discovering instrument, the shankus or pivots made of pancha-dhaaraa-loha or five alloy metal are the best.

Kshvinkaa, iron-pyrites, copper, indra, and ruruka, purified, powdered, and filled in mrugendra moosha crucible and boiled to 300 degree heat with beaked bellows, will yield a 5 alloy metal, strong and heavy.

Paara-granthika acid for insertion of the crystal, is described in "Moolikaarka prakaashikaa." Mercury, bamboo salt, Indian spikenard joint, paarvanika or clerodendrum phlomides, svarna seeds or Indian labernum seeds? or yellow thistle seeds?, and ghatotkaja or American aloe, in equal quantities, should be filled in a big bellied earthen pot, heated to yield a golden hued shining liquid, which is very useful for capturing reflections.

Chumbaka crystal is the one most suited for use in capturing reflections of objects. It is manufactured as per "Manipradeepikaa," with the following ingredients. Magnet, sand, borax, ivory, shoundika or long pepper, mercury, paarvana or clerodend rum phlomoides, copper, vermillion, iron-pyrites, grudhnika, souri or marking nut, buffalo hoop, vishwakapaala, cleaned and powdered, and filled in karpala crucible and baked in a furnace with the aid of owl-nosed bellows to 100 degrees, will yield a fine image producing crystal.

Pigment for coating the screen so as to present a clear picture, is called "Roopaakarshana-niryaasa," or image reproducing niryaasa or varnish. Out of 360 such varnishes that is the best.

Says "Niryaasa kalpa":

Moonstone, crownchaka, bamboo rice, five milks from banyan, fig, keg etc., trees, magnet, udusaara, mercury, mica, pearl, earth from ant-hill, saarasvata oil, and nakha or nail? these 16 articles to be taken in equal parts, purified, should be ground for a period of 30 days in the juice of the peacock's egg, then mixed with bilva oil and boiled for four yaamaas or twelve hours until it becomes a perfect gum or varnish. Some call it reflector varnish. Some call it virinchi-varnish.

The varnish is to be evenly spread on the special cloth called patadarpana, so that it may present as on a cinematograph screen, the pictures reflecting the location of anti-aircraft mines discovered by the roopaakarshana yantra.

The production of Pata-darpana is described in "Darpana-prakarana":

Gum, cotton, pratolikaa, kuranga or pallatory root, maatanga or keg tree bark, cowries, kshoneeraka, gholikachaapa, granite sand, parotikaa, sea-foam, priyangava, ghanjhotikaa, sugar-cane, rukma or argemone mexicana, kesara or mesua ferraa gum, earthen salt, suvarchala, urugha, bydaara oil, muchukunda flour, sinjaanu, anchaalika, turmeric, kaarmuka or acacia catechu, these ingredients in the proportion of 100, 58, 25, 28, 4, 12, 5, 3, 1, 30, 10, 5, 8, 12, 3, 13, 22, 27, 28, 3, 24, 7, 3, 13, should be cleansed, powdered, filled in a vessel, and boiled in the furnace with 100 degree heat, and the unified fluid should be poured on a flat surface so as to form an even surfaced sheet. After drying, the photographic niryaasa varnish is to be used to coat this sheet, for use in the Guhaa-garbha aadarsha-yantra.

Thamo yantra or Darkness creating yantra:

Vimaanaas are liable to be attacked by enemies with poison fumes of Rouhinee or krakachaarimani rays. As a protection against it the thamo yantra has to be installed in the vimaana. Out of 132 types of thamo-yantras, the 62nd variety is said to be the best for safe- guarding against poison fume and ray attacks by the enemy.

Black lead, aanjanika (collyrium?), vajra-tunda are to be powdered and mixed in equal quantities, filled in fish-shaped crucible and placed in crow shaped furnace, heated to 100 degrees, and poured into the cooling receptacle will yield a fine, light, strong thamo- garbha-loha, or darkness impregnated alloy metal, useful for making Thamo-yantra.

The peetha or stand is to be 3 feet wide and ½ foot high, square or round. In the centre of it is to be fixed the pivot. At its front should be placed the vessel of the acid of guggala or

Indian dellium. To the west should be fixed the mirror for en-
hancing darkness, and in the east should be fixed the solar ray
attracting tube. In the centre should be fixed the wire operating
wheel, and to its south should be fixed the main operating wheel
or switch.

Its working is as follows. On turning the wheel in the south
east, the two faced mirror fixed to the tube will revolve and collect the solar rays. By operating the wheel in the north west, the
acid in the vessel will begin functioning. By slightly moving the
wheel in the south- east, the solar rays will enter the crystal in
the acid vessel. By turning the wheel in the west, the darkness
intensifying mirror will begin to function. By turning the central wheel the rays attracted by the mirror will reach the crystal
and envelop it. Then the main wheel should be revolved with
great speed, when the darkness will be produced enveloping the
vimaana and making it invisible, and the efforts of the enemies
to attack it with poison gas and rays will miss their target and
become ineffective. This yantra should be placed in the north-
west sector of the Vimaana Panchavaataskandha-Naala.

Iron rust, shaarana, copper, suvarchala salt, in equal parts,
to be filled in mayookha crucible, placed in jumboo-mukha furnace, and using kaakamukha bellows boiled to 102 degrees and
cast in the yantra, will yield a pure, light, soft, strong, nice cool
metal known as vaatadhaarana loha.

4 tubes, each 2 yards long and 1 yard high, should be prepared. Like the circular opening in the top of the vimaana two
openings on each and one at the bottom should be prepared.
Each tube should be inserted in the said openings. Another tube
12 feet long and 3 feet high should be fixed on the western side
in the opening at the top. To each tube should be attached bellows' mouth operated by wheels. By turning the wheels of the
5 tubes the 5 poisonous winds will be sucked in and passed into
the tubes to make their exit, without causing harm to the plane.

Lohasarvasva says:

There are 13 air layers known as Vrishni and others. By the force of the Panktiraadhasa Kendra, they tend to jostle each other, and generate fierce forces which will be destructive to the unwary vimaana which may get involved in them. Therefore the Pancha-Vaata- Skandha-Naala Yantra is to be inserted in the back portion of the vimaana; to safeguard against evil consequences.

Roudree Darpana Mirror.

From the south-eastern side of the earth-sun axis solar rays touch the turbulent forces in the etherial regions, and burst into flames, and vimaanas which may be out on their course may be destroyed by the flames. To prevent such a happening the roudree-darpana yantra should be installed in the bottom of the vimaana.

Says "Yantrasarvasva", "At the time when spring passes into summer, the forces in the junctional regions of the sky, on contact by fierce solar rays, burst into tumultuous flames, and destroy all things that pass through, Therefore the roudree darpana should be fitted in the vimaana as a safe-guard against that."

According to "Darpana Prakarana," iron rust, magnet, veera iron, borax, panchaanana metal, mica, honey, red castor bark, banyan, suryavarchula or sweet-salt, gold, alika, shaarkara or benzoin tree bark, pancha tikta or 5 sours, snake gourd, and paaduka, are to be powdered, cleaned, and in equal quantities filled in padmaasya crucible, and placed in vishvodara furnace and heated to 200 degrees. The molten liquid poured into the mould will yield excellent flame-proof roudree-darpana glass.

With this roudree-darpana glass a plank of 16 feet in dimension should be prepared. A pivot 25 inches thick should be fixed at the centre of the plank. At the edge Of the pivot, two wheels should be fixed revolving with right motion and reverse motion for expanding and contracting. A wheel equipped with rods for spokes should be fixed, the spokes being 15 inches from each other. Sheets made of roudree glass, washed with linseed, drona or lucas aspera, liquid amber, and madder root oils should be fixed to the rods with hinges. Similarly crystals made of roudree-darpana glass, with 5 facets, cleaned with the oils should be fixed at the end of the rods. Between each rod 18 leaves like lotus leaves with revolving keys should be fixed. The instrument is to be shaped like an umbrella. The leaves should be fixed at the pivot top with 8 keys.

When the burning flames are imminent, the pilot should turn the expansion wheel vigorously, and the umbrella will open up and provide a shielding cover for the vimaana. The lotus petals, the crystals, and the enveloping cover will protect the vimaana from the threatened danger.

Next, the Vaata-skandhana-naala.

According to "Gati-nirnaya-adhyaaya"

In the Aavaha and other giant wind spheres there are 122 kinds of different motions of the wind. In the summer season the 79th kind of motion occurs mostly. When the vimaana travels in the 4th region of the sky, it tends to zig-zag owing to the wind currents, and cause hardship to pilots and other occupants. Therefore as a safe-guard against it, the Vaatastambhana-naalayantra should be installed in the bottom section of the vimaana.

Says "Yantra Sarvasva",

The vaatastambhana naala yantra should be manufactured with the vaatastambhana metal only. According to "Lohatatva prakarana," dantee or croton seeds, suvarchala or sun-flower salt, mayoora or sulphur, lohapanchaka or copper, brass, tin, lead, and iron, bhrisundika, suranjika or sulphate of mercury, varaahaanghri loha, virohina or creya arboria, kuberaka, muraari-kaanghri metal, ranjika or phosphorus, suhamsanetraka, dala or folia malabathy, courie sea-shell, mrinaalikaa or lotus stalk, to be powdered, cleaned, and in equal quantities filled in matsya or fish-shaped crucible, and placed in maaghima furnace, and with the aid of vijrimbhana bellows duly melted, will yield a molten liquid which when poured into the mould and cooled will yield an excellent vaatastambhana loha.

With that metal 6 naalas or tubes of 15 inches diameter, with wide openings should be prepared and fixed in the tail and centre and front of the vimaana 10 inches deep, east to west and north to south, and held together with hoop iron binders. At the mouth of each tube a vaatapaa or air imbibing crystal should be fixed by wires. Between the tubes flags or pennants made of cotton-cloth duly processed, should be tied. And wheels made of the special metal should be fixed above each pennant. When the vaataayanee wind blast blows, the pennants will flutter noisily, and the wheels fixed underneath them will also revolve as also the crystals. The fluttering pennants pass the blowing wind to the wheels which pass them on to the crystals, which will pass them into the tubes from which they will be ejected through openings to the outside. That will protect the vimaana from their interference.

Next Vidyuddarpana Yantra.

Sowdaaminee kalaa explains it as follows:

During the rainy season, when rain clouds gather in the sky, lightning of five kinds begin to play. They are named vaaruni, agnimukha, danda, mahat, raavanika. Of them, vaaruni and ag-nimukha are very active and fearful and are likely to be attracted by the roudree- darpana and other mirrors and cause fires which destroy the vimaana. In order to prevent that the vidyud-yantras should be installed in the front and the right side of the aeroplane.

According to Yantra-Sarvasva:

In order to protect the plane from lightning, vaaruni and agni, the vidyuddarpana should be installed in the vimaana.

Darpana Prakarana explains its structure:

Kuranga or pallatory root, panchaasya metal, virinchi, shona-ja or red lead, sand, alum, kutbha or hellebore, pearl, sundaaliga,

mercury, yavakshaara or salt-petre, borax, bidouja salt, pingaak-
sha or terminalia chebula (?), cowries, and karbura or hedychum
specatum, powdered and purified, in the proportion of 10, 7, 4,
3, 12, 2, 3, 7, 11, 27, 14, 3, 22, 18, 5, and 11, filled in padmaasya
crucible, placed in vishvodara furnace, and with the aid of the 5
mouthed bellows heated to 500 degrees, the molten liquid will
yield in the cooling mould a glass which is impregnated with 300
shaktis or forces, and can overpower the lightning blasts from
the vaaruni and agni forces, shining with wonderful rays, and
capable of spreading its own lighting force within 2 kshanas or
a few seconds to a distance of 5 yojanas or 15 miles.

With that lightning darpana glass should be constructed the
Vidyuddarpana yantra. A plank, 20 feet in diameter and 1 foot
high, square or circular in shape should be prepared, 4 glass tubes
of crescent moon shape should be fixed around the peetha or
plank. In the centre should be fixed a cage made of chumbuka
glass, fitted with wires and 5 faced switches at each face, and 5
goblets made of vidyuddarpana. In the centre should be fixed
a spire made of the same glass with 7 cross spokes and tubes,
8 faced and 10 angled. By turning the key inside, the spire is to
revolve with speed. That will attract and contain the lightning
emitted by the clouds. The rays will expel it to the outer air
region, and incapacitate it. Then a snow-like cool temperature
will render the interior of the vimaana safe and pleasant for the
pilot and other occupants. Therefore this vidyuddarpana yantra
should be installed duly in the vimaana.

Shabda-Kendra Mukha Yantra.

"Kriyaasaara'" says—

The spots from which sounds emanate in the sky are called
shabda-kendras or sound centres. The different directions
from which the sounds are projected are called shabda-kendra
mukhas. The yantra which is meant to control the sounds so
projected is called shabda- kendra mukha yantra.

Out of 304 classified sounds the sounds of water-laden
clouds, wind, and lightning are said to be fiercest. In the 8th

region of the sky these three sounds unify in the shishira Ritu or February-March period, and produce ear-splitting thunders. They would deafen pilots and others in the vimaana. As protection against that the shabda-kendra mukha yantra is to be installed.

It is said in Shabda-nibandhana, "By the combination of water, fire, air, and sky, sound is generated both among living and life-less objects. The sounds in the word "Shabdaha," i.e., sha, b, d, and ha, indicate water, fire air, and sky symbolically."

"Naamarthha-kalpa" says,

We shall deal with the nature of sound or "shabdaha". The word consisting of sounds sha, ba, da, and ha, stands for water, fire, air, and sky. By the combination of these four forces in various proportions, 304 different kinds of sounds are generated.

The Braahmana bhaaga of the Veda also says shabdaas are of 304 kinds, such as sphota or embryo, very feeble sound, feeble sound, manda or soft, very soft, fast, very fast, medium, very medium, great sound, thunder sound, and thunder-bolt sound.

It is said in Yantra-Sarvasva,

In the 8th region of the sky, by the concatenation of watercloud, wind and fire, an extremely fearful thunder clap will occur which will blast the ears of pilots who may enter the region. To safeguard against that the vaataskandha mukha yantra is to be installed in the vimaana.

In the 8th region of the sky there are 307 centres of sound. From the 70th centre a fierce sound proceeds by the force of water. From the 312th centre a fierce sound produced by wind will emanate. Similarly from the 82nd centre a fierce sound generated by lightning will emanate. By collision of the three a terrible sound will result which will deafen the pilots of the vimaana. Therefore facing each sound emanation centre the shabdopasamhaara yantra is to be established.

The construction of the yantra is as follows:

Gavyaarika, monkey's skin, duck-weed, shana-kosha or jute product, crounchika or lotus stalk, vaaripishtaka or shag, roonthaaka, flesh, elephant trunk, and tin, are to be purified, and the nine elements, other than the skin, in equal parts, filled in niryaasa yantra and baked for 3 days with buffalo bile, will yield a decoction of fine scarlet colour. Seven times this decoction should be spread on the skin, and left to dry in the sun. The skin will then acquire the capacity to suppress sound.

A box 2 feet long and 1 foot high made of badhira or deaf metal is to be made. Two pipes made of the same metal shaped like crane's beak, should be fixed inside it. Above it should be fixed an umbrella made of shabdapaa darpana, or sound-drinking glass. A crystal washed with tulasee or basil seed oil should be placed inside the monkey skin and sealed with rhinoceros gum. The sealed skin with crystal should be placed in the central pipe inside the box. Monkey skin alone should be placed in the pipe on the left side. Thin wires should connect them all and be fitted with hinges and switches. Above the canopy of the box a monkey skin shaped like lion's mouth should be connected by wire through a pipe to the crystal in the tube inside the box. The top of the box should be covered, securely.

Badhira loha or deaf-proof metal is explained in Lohatantra-Prakarana; lime fruit, laguda or sweet-scented oleander, virinchi, rishika or water-calteop, maaloora or Bengal quince, panchaanana metal, luntaaka, varasimhika or solenum xunthokurpum, kuravaka or gigantic swallow-wort, sarpaasya or mesua ferrea, vaakula or surinam medlar, jack-fruit, camphor and vatika or salvinia cusullata, in equal parts, purified, and filled in tryutee crucible, and heated in the furnace, will when cast produce a metal, cold, dark, sound-proof, powerful, able to control bleeding, and draw out missile parts from the war wounds of soldiers and healing them, and capable of reducing the effect of thunder claps.

The simhaasya bellows is to collect the fierce sound and transmit it to the crystal inside the metal box so that the monkey skin will absorb it and stifle its intensity. Therefore shabda-kendra mukha yantra should be installed in the vimaana.

Vidyud-dwaadashaka Yantra, or Yantra of 12 lightnings is explained in Kriyaasaara.

In the realm of the comets and shooting stars in the sky, at the 8th region there are 30703221 shooting stars. 8000 of them are prone to lightnings, and 12 of them known as mahaakaala etc., are of importance.

Shakti-tantra says, "The 12 lightnings which form the eyes of the shooting stars are named, rochishee, daahakaa, simhee, patanga, kaalanemikaa, lataa, vrindaa, rataa, chandee, mahor-mee, paarvanee, mridaa."

Kheta-sarvasva Says:

Mahaakaala, mahaagraasa, mahaajwaalaamukha, visphulinga mukha, deerghavaala, khanja, mahormika, sphulinga-vamana, ganda, deergha-jihva, duronaka, and sarpaasya are 12 comets with 12 lightning eyes.

The lightning effects of the comets are extremely severe in the period of sharat or autumn, October and November, and vasanta or spring, March and April. By the collision of the solar rays and the lightnings a force called ajagara is created. When the vimaana reaches the 20th region of the sky, that force paral-yses the plane. To protect against such happenings the vidyud-dwaadasha yantra is to be installed.

Yantra Sarvasva also Says:

Vidyuddwaadasa yantra is excellent in protecting against the lightning effect of comets. Its details are as follow. First duly coated jataghana should be prepared. It should have 22 folds so as to cover the vimaana. Poundraka and other crystals should be fixed in each of the folds. Then mahorna acid should be placed inside in the north-east side of the vimaana. 8 rods, each of 6 arms length, made of anti-lightning glass should be fixed in the 8 directions over the cover of the vimaana. At the beginning, middle, and end of the vimaana canopy, spring wheels made of dambholee metal, 5 faced and interconnected should be fixed with revolving bolts. Cages made of wire should enclose the poundraka crystals, and the wire terminals should be attached to the spring wheels. The wire ends from 4 of the cages should have a common switch.

On the main wheel being put in speedy motion the 12 crystal cages will revolve, the enveloping cloth cover will spread out, and the lightning absorbing power of the cloth will be activated. The crystals will attract the ajagara lightning, bifurcate the comet lightning from the solar rays, and transmit it to the 8 rods. The rods will absorb and then transfer the lightning power to the folds of the power proof cloth. By operating the central switch in the enclosure, a force called vidyut-kuthaarika, or lightning-axe, will be generated in the acid, and it will attract the comet force from the cloth, and submerge it in the acid. Then by operating the end switch in the enclosure, the ajagara force in the acid will dart towards the pataghana cloth-cover and take refuge, where upon the blowing wind will evaporate it and nullify its effects, and the vimaana will be out of danger.

According to Darpana Prakarana,

Shundaala metal, mridakaantaka or mountain ebony, ghanodara, budilaakara or tamarind, vatsanaabha poison, pankaja or eclipta prostrala, kutilaraga, naga or mesua ferrea, white sand, vara or syndhava salt, garada, mica, garala, or honey product, mukha, shringa, sphatika crystal, avara, muktaaphala or pearl? guggulu or boswellia glabra, kaanta or steel, kuranja or Indian beach, natron, salt-petre, borax, copper, snake scale, udupa, barren tree, sonamukhee or Tinnevelly senna, brown barked acacia, jaambalika or citrus grass? lemongrass? kusha grass, kudmala or flower bud, gold, these 26 ingredients, purified and filled in crucible and placed in padmaakara furnace, and with the aid of simhaasya or lion-faced bellows heated to 300 degrees, and poured into the mould, will yield a fine anti-lightning glass.

Dambhola loha or thunderbolt metal is thus described in Lohatatitra-Prakarana:

Urvaaraka, kaaravika, kuranga, shundaalika, chandramukha, virancha, kraantodara, yaalika, simhavaktra, jyotsnaakara, kshwinka, pancha-mourtwika, metals should be purified and placed in mandooka-or frog-crucible, placed in the five faced furnace, and with the aid panchamukha or 5 mouthed bellows heated to 500 degrees, will yield the dambholi alloy.

Poundrika crystals are described in Maniprakarana which describes the poundrika crystal.

Poundrika, jrimbhaka, shibira, apalochana, chapalaghna, amshupamani, veeragha, gajatundika, taaraa mukha, maandalika, panchaasya, amrita sechaka, these 12 crystals are destroyers of ajigara.

Draavaka prakarana explains mahorna acid: pynaaka, panchamukha, ammonium chloride, wild liquorice, iron-pyrites, kudupa, vajrakanda, budila, mercury, steel, charcoal, mica, these in equal parts purified and boiled in acid boiler, will yield mahorna acid.

Praana-Kundalinee Yantra.

According to "Kheta-Sarvasva," where the contact of smoke, lightning and wind courses in the sky occurs is the praanakundalee position. The yantra which can control, restrain, and set in motion the forces of these three in their several courses, is called praana- kundalee yantra.

According to Kriyaasaara, the yantra which is installed in the praanakundalee kendra of the vimaana in order to control the forces of lightning, wind, and smoke, and adjusts their movements is called praana-kundalinee yantra.

Says Yantra-sarvasva, "In order to control the movements of the forces of smoke, electricity and wind, and make them disperse, move, halt, or make stunt move or reverse move, the praana-kundalinee yantra is installed in the vimaana. A peetha or stand 3 feet in diametre and 3 feet high, square or. circular, should be made of vrishala metal, with 8 kendras or central spots. In each central spot, two wheels with revolving hinges; small peethaas or plates with 3 holes, 4 teeth, 3 pivots, in their middle a central pivot, three red- coloured tubes or pipes with opening and closing wheels, and switches for right motion, and reverse motion, with a shabda-naala in the centre, with wheels (with hinges and rods) which will flap the wings; from the north-east and south-east kendras and the middle- kendra in the west up to the middle of the course of the yaana kundalinee revolving wheels with pivots. Motions are by means of hand wheels. By the operation of the several wheels the plane will be set in motion. From the central pivot of the 8 kendras strong wires should pass the eastern peetha or footboard through randhras or holes and reach the tops of the 3 tubes at the window. The 3 forces should be made to aid the motion of the vimaana, and the remnant of the force should be passed through the 8 tubes and get lost in the sky, leaving the vimaana unperturbed."

Shaktyudgama Yantra

The eight powers of the planets and stars, at the time of full moon in the month of kaartika,—i.e., November-December, are pulled forcefully by mahaa-vaarunee shakti or great cold force. In the 137th route in the sky there is a jala-pinjooshikaa shakti which will attract and spread them all over, and there will be a fierce outburst of dew and snow. Then 3 currents will be generated: one will be a damp cold air flow; the second will be a wet dewy flow; and the third will be a cold air flow. When the vimaana approaches that region, the first force will divest it of all power. The second force will benumb the pilots and operators. The third force will envelop the vimaana and make it invisible. Thus overcome, the vimaana will crash. As protection against such a happening the shaktyudgama yantra should be fixed in the navel spot of the vimaana.

"Khetasangraha" says,

"Eight planets are, Mars, Sun, Saturn, Venus, Mercury, Moon, Jupiter, and Ruru. And krittikaa, shatabhisha, makha, mrigashiras, chitra, shravana, pushya, and ashvinee are eight luminous stars. In the course of their transit through space the planets and stars approach each other in the period of sharat or autumn generating eight forces."

"Chaara-nibandhana" also says, "According to the science of astronomy, planets and stars in the course of their perambulations happen to approach one-another. Then conflict arises between the magnetic and electric forces of planet and star, and eight cold forces are generated in consequence."

"Shakti-sarvasva" says, "When the star krittika comes near planet Mars, a force called shaktyudgamaa is generated. Similarly, by the star shatabhisha coming near the planet, Sun, a cold force called jwaalaamukhee is generated. By the nearness of mrigashiraa and Venus a damp cold windy force called mahojjwalaa is generated. By the approach of star makhaa to planet Saturn a force called shytya-damshtraa is generated. By the approach of chitta to Mercury a force called shytya-hymaa is generated. By the approach of shravana to Moon a cold wave force called sphoranee is generated. By the nearness of pushya to Ruru a force called mahormilaa in generated. And by the approach of ashwinee towards Jupiter a force called mandookinee is generated.

These eight forces, shytyodgamaa, sheeta jwaalaa-mukhee, shytya-damshtraa, sheetarasa-jwaalaa, shytya hemaa, sphoranee, sheetarasa-ghanaatmikaa, and shytya-mandookinee, by mutual inter-play according to the seasons, will become six new forces."

Says "Ritukalpa,"—"In spring the differing forces will be 5, in summer 7. in the rainy season 8, in autumn 3, in hemanta or cold season 10, and in winter 2.

The 3 forces during autumn are as follows. The planet-star forces by contact with the sun's rays, assume 3 forms. Sheeta-jwaala, shytya-damshtraa, and shytyodgamaa, coalesce and become sheetarasa-vaata shakti. Shytya-rasa-jwaalaa, shytya hymaa, and sphoranee coalesce and become vaari sheetasheekaraa shakti. Shytya-ghanarasaa and shytya mandookinee become sheeta-vaata- rasa-praavaahika shakti."

Yantra Sarvasva Says,

"To protect the vimaana from the effects of these three forces the shaktyudgama yantra should be installed.

First with the shytyagraahaka loha or cold-absorbing metal, protective hoods should be prepared both for the pilot and for the plane. At the front and tail portion of the aeroplane cover should be fitted switches for contraction and expansion. In the front or elbow hinge of the supporting beam of the covers the two sandhi-keelie should be fitted. Three tubes should be prepared out of the cold-proof glass, and should be fitted in front and on either side of the pilot's cock-pit. Bhraamanee chakra or wheel should be fitted at the front. When the three shaktis or forces attack the vimaana, the expansion wheel should be revolved vigorously. It will first cover the pilots and then cover the entire plane also. By operating the bhraamanee wheel the attacking forces will be slowly absorbed, and the shaktis will be forced through the cold air tubes. By operating the main switch of the naala tantries, or wires, the forces will be made to go through the tubes into the outside air, and vanish therein. The pilot and the vimaana will both be saved from danger.

"Loha-tantra" describes the shytya-graahaka loha, or cold absorbing metal as follows:

Blue lotus, crowdika or rhinoceros horn or vaaraahi root, somakanda, vishwaavasu, crownchika alloy, chandrakaanta or moon-stone, vaardhyashvaka alloy, varuna tree, 5 kudmalas, simhaasya, shankhalavaa, and goose-berry, to be purified and in equal quantities filled in shundaalaka crucible, placed in chanchoomukha kunda, and with panchaanana bellows heated will yield a fine cold-capturing alloy.

"Darpana prakarana" describes cold-proof glass: lead, kapaalee, moonstone, castor, margosa seed, trinaanga or cus-cus grass, kshaara-traya or natron, salt-petre, and borax, suvarchalaa or sun-flower?, fine sand, bhaarika, collyrium or eye-black, kuranga or pallatory root, panchormikaa, chandrarasa, and shivarika, purified and in equal quantities filled in simhika crucible, placed in padmaakara furnace, and with the aid of shoorpodara bellows heated to 300 degrees, and poured into mould and cooled, will yield an excellent sheetaghna darpana or cold-proof glass.

Vakra-prasaarana Yantra:

Enemies attempt to destroy one's vimaana by missiles and dambholi and other mechanisms. The pilot should discover them by means of mukura and other yantras and immediately change course and avoid the trouble. Therefore the Vakra-prasaarana yantra, or diversion enabling mechanism should be installed in the vimaana.

Yantra Sarvasva says,

"When there is danger from dambholi and 8 other kinds of destructive mechanisms contrived by enemies, in order to escape that danger the vakraprasaarana yantra is prescribed:

Sulphate of iron, sacred peepul gum, and copper 16 parts, krishnaaguru or black sandalwood 3 parts, zinc 5 parts, collyrium 1 part, should be purified and mixed and boiled with 100 degree heat. Aaraara copper alloy will be formed, goldish and light and hard. A wheel 3 feet wide and 3 feet high should be made out of it. It should have a pivot, and be installed in the bottom of the eeshaadanda axle moola of the vimaana. Four inches thick and of arm's length, with 16 wheels having band-saw toothed edges attached to two pivots, oil-cleaned, with 3 joints, with oil-cleaned rods attached to the saw-toothed wheels, with keys; in the middle should be fixed 2 keys which will eject smoke, and 2 keys which will shut off smoke. Proper wiring should connect the several parts. This will enable the vimaana to zigzag like a serpent, to reverse, and to divert so as to avoid the danger zone, and get out safely."

Shakti Panjarakeela Yantra:

In order to provide electric force to all parts of the vimaana and make them operate smoothly the shakti-panjara-keela yantra is to be installed.

According to Yantrasarvasva, "As a means of charging all parts of the vimaana with electric current the shakti-panjara-keela is prescribed. It is made as follows:

Steel, crownchika alloy, and iron, in the proportion of 10, 8, and 9, to be powdered and filled in crucible, and placed in aatapa furnace and heated to 100 degrees and charged with 10 degrees of electric current, will yield Shakti-garbha metal with which the yantra is to be made.

A peetha or plank of arm's length and equal height, should be made out of above metal. In the middle and at either end of the peetha three pivots with half moon shaped hinges should be fixed. A flat bar made of copper should be fixed and tightened with bolts. Pipes with holes are to be made out of the metal and equipped with rods fixed in the holes, and connected all round with wires, forming a strong caged globe. The cage should be fixed at the top of the copper band. For the rods and wires in the cage to receive electric current a switch should be duly fixed at the bottom of the cage. And switches should be provided for all the 32 parts of the vimaana for electric connection and disconnection. This enables the plane to career through the sky in any desired direction."

Shirah-Keelaka Yantra.

It is said in Kriyaa-saara, "When the plane is passing through a region of overhanging clouds, there is possibility of lightning striking and destroying the plane. As protection against that the shirah-keelaka yantra should be installed at the crest of the vimaana."

The Yantra is described in Yantra Sarvasva:

"When there is danger of lightning striking the plane, the shirah-keelaka yantra is to protect it. Therefore it is explained below. An umbrella, of the same size as the top of the vimaana, with ribs and metal covering should be made out of vishakantha metal. The umbrella stick, of arm's length, and peetha or stand, circular in shape, should be made out of the same metal. Then out of baka-tundila metal three wheeled keelakaas or hinges should be fixed at the front, back, and middle of the vimaana. The umbrella rod should be fixed in the middle of two keelakas.

The agnikuthaara crystal with metal cage should be fixed at the top like a crown. A three wheel switch revolving key should be fixed by the side of the pilot. Then wires made of kulishadh-wamsa metal should be run from the crystal to the three wheeled revolving keelaka. In front of it shabda-naala tube with switch should be fixed. The yantra should be enclosed in a cover made of suranjikaa glass. When there is anticipatory thunder in the clouds, the glass covering cracks, and the tube of the wiring will emit sounds, and the wires will be severely shaken. When the pilot notices these signs, he should quickly set in motion the three-wheel keelee, which will revolve the umbrella with 100 linka speed. Then the crystal switch should also be turned, where-upon the crystal will also revolve with intense speed. By the speed of the revolution of the umbrella, the force of lightning will be stemmed, and the danger will be passed, leaving the vimaana and the pilot safe. That is the use of the shirah-keela yantra."

Shabdaakarshana yantra:

In order to tap or discover the sounds in the 8 directions of the vimaana, wired or wireless, up to 12 krosas or 27 miles, caused by birds or quadrupeds or by men, with 8 mechanisms, the shabdaakarshana yantra is prescribed to be fixed in the shoulder of the vimaana. A peetha or foot-plate four-cornered or round should be made out of bidaala metal, with a pivot in the centre. On either side should be fitted machines which will attract any kind of sound and repeat it. With the soft leather of roruva or grinjinee bird two ball-shaped domes should be fixed. Between them in a suraghaadarsa vessel katana-drava acid should be filled and the vessel should be installed. Above the acid vessel and between the two globes should be fixed sound spotting rod made of sound capturing ghantaara metal, fitted with a bunch of wires. It should be enclosed in a cover made of kwanaka glass. In the corner three thumb size wheeled knots should be fixed. From them to the rod fine strong wires should be connected. Enclosing the wires a karanda or container made by kwanaadarsa glass, with small holes should be placed. A vessel made of the same glass, shaped like a drona or grain measurer, should be placed on top of it. In the east and west and north and south 4 crystals named rudantee-ratikaa should be arranged with wires. Above it shabdaphenaka covering, with small shankus or screws fixed, should be placed. It should be covered by a covering made of kwanaadarsa glass, with 8 small holes. Wires starting from the screws and passing through the holes should reach the top of the covering. In the centre of it in an inch size hole simhaasya-danda-naala or tube should be fixed. In front of it a vaataapaakarshana chakra or wind wheel with 16 spokes with wires should be fixed. The wheels should be fixed in all 8 directions. In the simhaasya mukha naala or lion-faced tube on 8 sides revolving wheels should be fixed. 8 goblet like vessels made of pure vajeemukha metal should be fixed. Wires from the 8 holes of the covering should be placed in them. Similarly from the wind wheel wires should be connected to 8 screws in the 8 goblets on the simhaasya. Then from the 8 screws in shabda-phena, wires should be connected to the crystal in the acid vessel.

By the flow of wind the wheel turns with speedy right and left motion, and will set in motion the shabda-phena wheel. Then the wheels on the 8 screws also will turn. Then the sound detection rod made of sound-capturing ghantaara metal will be set in motion. Thereupon the two globes made of roruva-grinjinee skin will attract all sounds clearly and store inside themselves. By moving the central switch there the sounds will pass through the simhaasya tube and enter the dronaasya vessel, and make the sounds clearly audible to the hearer. The pilot will listen to the voices and direct the plane away from the vicinity of danger. Hence the shabdaakarshana yantra is prescribed.

This yantra is in 32 varieties. And it is distinct from the 32 parts constituting the vimaana.

Of the materials required for this yantra,—Byndaala Metal, according to Lohasarvasva, is made as follows:

Zinc, sharkara or quartz powder?, kaanta or steel, mica, shilaarasa, kamatha or benzoin, dimbhaari, areca-nut, karagrathinee, copper, virinchi, karna or sal tree, patalee or long blue cucumber, gumbhalee, dumbholika alloy, kshaara or chloride, kraantika, simha, panja or momordika, dalinee, mercury, eye-black powder or surma?, kshonika, veera or red-lead, yellow thistle, madder-root, mridarutee, brass, iron, these articles should be powdered, and purified in equal quantities, filled in shashamoosha crucible, placed in mandooka furnace, and with five-mouthed bellows heated to 200 degrees and melted to eye-level, when cast will yield a fine, light, blue, bydaala alloy.

Rutana acid is explained in "Moolikaarka prakaasikaa" as follows:

Yellow thistle, karanda or iron pyrites, wild liquorice, paarvani or chlorodendrum phlomaides, chanchooli or red castor, bhantikaa or madder root, kaarambha, vishwesha, chandikaa or sesbenia grandiflora, amara or Indian turnsole?, shundaalika, barbaraasya, sowrambha or tooth-ache tree?, praana-kshaara or ammonium chloride, virinchi, borax, arka or calotropis gigantia, surubhee or basil, these in the proportion of 4, 3, 3, 5, 7, 12, 15, 1, 3, 10, 24, 25, 30, 12, 20, 8, purified and filled in vessel and boiled to 108 degrees, will become a fine yellow rutana-draavaka acid.

Ghantaarava metal is explained in Lohatantra:

Bell-metal, aaraara, ruchaka or patron.?, gaaruda or emerald stone?, shalyakrintana, panchaasya, veerana, rukma or gold-metal, shukatunda, and sulochana, these 10 metals purified and powdered in the proportion of 5, 3, 12, 2, 3, 7, 5, 30, 4, 24, should be filled in shukti crucible, wrapped all round with earth, placed in alaabu shaped furnace, and boiled to 500 degrees up to eye level, should be poured into the mould. A fine, light, scarlet metal which will record all sounds will result.

Kwanadarpana mirror is explained in Darpana Prakarana:

Wild liquorice seeds, red catechu, false catechu, white catechu, garadaka or a poison, 8 kinds of salt, salyaaka, vara or sodium chloride, sharkaraa or granite powder, budilaka salt, jwaalaamukha or wolf's bane?, tundila or kayidonda, bydaala or arsenic?, shukatunda, ravimukha or magnifying glass, chancholika or red castor seed, arjuna or tin, luntaaka, varataala or yellow orpiment?, kuravaka or crimson thorny amaranth, kambodara, kaamuka or punnaaga or Alxandrian laurel or pinnay oil tree, these ingredients, after triple cleaning, are to be filled in padma crucible, placed in padmaakara furnace, and heated to 700 degrees, and poured in mould, will yield an excellent kwanadarpana glass.

Rudantee-mani is explained in "Mani Prakarana":

Kshaaratraya or natron, salt-pare, borax, aanjanika or eye-black powder, kaanta or sun crystal, sajjeeka, vara or sodium chloride, karna or oxide of arsenic, cowrie shells, maakshika or iron pyrites, sharkara or granite grains, sphaatika or alum, kaamsya or bell- metal, mercury, taalakasatva or yellow orpiment, gyra or marking nut, ruruka, rouchyaka, kudupa, garada or aconite, panchamukha metal, shingara or iron dross, and shundolika or great leaved caledium, these 21 articles, purified, and filled in aanika crucible, placed in shouktika furnace and boiled to 103 degrees, and cast into maniyantra mould, becomes a fine rudantee crystal.

Ruchika mani also is explained in Maniprakarana:

Sea-foam, chamaree cat's nail and mouth bones, steel, paarthiva, granite grains, shilaarasa or liquid amber, mercury, praana- kshaara or ammonium chloride, alum, naaga, cowrie, maakshika or iron pyrites, shundaala or great-leaved caledium, rundaka or eagle wood, kudupa, suvarchala or natron, jambaalika, musk-cat's tooth, or yellow orpiment?, ranjaka, manjishtha or madder root, paarvani or stag-horn, rukma or gold quartz, yellow thistle, owl's nails, vara or ammonium chloride, oyster shell, these ingredients, purified and filled in equal proportion in nakhamukha crucible, placed in mahodara furnace and heated with the aid of six-faced bellows to eye-level, and poured into mould will yield a strong, dark, heavy rutikaa crystal.

Shabda phena mani is described in "Shabda-Mahodadhi":

"Take badaba sound from the sky, life-giving trait from water, the fire of air from the atmosphere, the echoing quality from boulder, the splitting quality from solar-rays, moss layer, sea-foam, bamboo, conch; manjishtha or madder root, kusha grass, gribhdnaka, rudra- shalya, gokarna or sal?, and musali or curculigo orchioides, in the proportion of 7, 22, 45, 13, 32, 19, 38, 14, 22, 38, 42, 13, 25, 9, and 23. These purified and boiled will yield shabdaphena."

From moss-layer to musali the ingredients should be purified, and in the said proportions should be filled in phenaakara yantra, should be baked for 3 days, and for a week the sankalana key-wheel of the yantra should be turned in full speed for half a ghatika daily, when foam will be formed. The foam should be filled in shakti-sammelana yantra. Then through 6 tubes the 6 shaktis or powers from praanana to sphotana should be injected into the foam patiently. On either side of the yantra switches should turn the mixing or churning wheel inside the yantra. Then moderate heat should be applied from praanana to sphotana power infusion. Then keeping it in the sun, electric power should be applied to the foam up to 85 degrees. This electric cooking should be done for 6 days. Then carefully extracting the foam from the yantra, it should be stored in the vaajeemukha metal box. That shabda- phena would be able to attract and record all kinds of sounds.

Vaajeemukha metal is described in "Lohatantra":

Copper 3 parts, sonamukhee or iron pyrites 2 parts, zinc 8 parts, veera or black metal 2 parts, kaanta or steel 3 parts, bambhaarika 1 part, kamsaarika 3 parts, panchaanana 6 parts, gowreemukha or mica? 2 parts, shundaalaka 6 parts, these 10 articles to be purified and filled in shundaalaka crucible, placed in shoorpaasya furnace and heated with vajraanana bellows and poured in vajraanana yantra and churned energetically for proper cohesion of the liquid, will yield vaajeemukha loha of light reddish brown colour.

Pataprasaarana yantra is described in Kriyaasaara:

In order to realise dangers to the vimaana en route, and shift directions towards safety, pataprasaarana yantra is prescribed. Says Patakalpa:

"Munja grass, lac, sal, red brinjal, shaambaree or arjuna tree bark, jute, raajaavarta or sphatikaari or hydrorgirum sulphuratum, darbha grass, kravyaada or Indian spikenard, with triple purification, and thrice exposing to soorya-puta or sun-baking, placing them in the cooking vessel, and baking for 3 days. Then the product should be filled in kuttinee yantra, and churned for 3 yaamaas or 3/8 of a day, then placed in cooking vessel and rebaked for 3 days. Then it must be poured into patakriyaa yantra or cloth-forming machine and churned, so as to form an even emulsion, and that will form a fine artificial cloth. It should then be coloured with seven colours. It should be rolled round a long pole, and the pole fixed in thrimukhee-naala yantra, and equipped with a key should be installed in the shoulder of the vimaana.

When the flag-like contraption shows red indicating danger ahead the pilot should loose height and reach safety. When favourable colours are shown, the pilot should note their significance and move the vimaana in the favourable direction.

Dishaampati Yantra: says "Kriyaasaara,"

"In its passage in the sky in the eight directions, the vimaana is likely to meet 15 fierce hurricanes called kowbera by the effects of the planetary forces With solar rays and unfavourable

seasonal conditions. They will cause baneful skin effects on the
occupants of planes and throat and lung troubles. To protect
against that the Disaampti yantra is to be installed in the left
shoulder of the vimaana."

The yantra is described in "Yantra Prakarana":

"In order to act as antidote to the poisonous effects of the
kowbera winds, F shall describe the disaampati yantra. A peetha
or foot plate, quadrangular or circular, should be made of paarvanee wood cured three times with requisite acids."

Paarvanee wood is described in "Agatatwa Laharee":

"Parvanee wood is wood which has very close joints as in
sugarcane. It is red coloured, long leaved, decked with red flowers. It has small thorns, is antidote for snake-poison, is acrid in
taste, and is used in driving away demons and other evil forces.
It blooms in the dark half of the month."

In the centre of the peetha, a tube or pivot made of the
19th type of glass, with 9 holes, 9 switches, and 9 wires, and of
arm's length, should be fixed. Eight kendras or centres should
be spotted in its 8 directions. 8 naalas, pipes, or tubes, should be
made 2 feet long, 6 feet high and 3 feet wide, and round in the
middle. A lotus with 8 petals should be prepared and fixed on
the top of the pivot. The whole should be covered with hare-
skin. Manchoolika linen should cover its mukha or entrance.
The wires in the tubes should be taken to the petals above the
lotus and fixed in the joints.

Manchoolikaa linen is explained in "Pata-pradeepikaa":

Vaasantee or gaertnera racemosa creeper, mrida, ranjikaa or
betel or madder root, ruchikaa or citron or castor, samvartakee
or myrabalan belliriki, phaalgunee or sepistan plum, chanchora
or red castor, arunakaanta or sun-flower, kudalinee, mandoorika
or iron dross, maarikaa or cubed pepper, lankaari, kapivallaree
or elephant pepper, vishadharaa, samvaalikaa, manjaree or asho-

ka, rukmaangaa or cassia fistula, dhundikaa or acacia sirisa, arka or gigantic swallow wort or madar, garudaa or coculus cordifolious, gunjaa or wild liquorice, and janjharaa.

Taking the twigs, shoots, leaves, buds, tendrils and barks of the above ingredients, and putting them in the baking vessel, they should be well cooked. Then add crowncha acid and boil again for 3 days. That will produce a soft white, pure, strong, fine manjoolikaa linen cloth.

Vaatapaa crystal should be placed in it. Amshupaa mirror should be placed at its front. When the Kowbera whirlwind contacts solar rays, then the amshupaa mirror will show a red and blue tinge. Then the keys of the nine tubes should be turned with great speed. By this a force will be generated in each tube, and passing to the hare-skin, a strong force called sammarshtikaa will be generated. The manjoolika cloth will receive that force and pass it to the lotus petals, and the petals by means of wires will transmit the force to vaatapaa crystal. The crystal, will, with the aid of sammarshtika force, absorb the Kowbera evil wind and throw it out through the lotus petal tube to merge in the outside atmosphere, and no harm will be caused to the occupants of the vimaana. Therefore disaampati yantra should be installed in the vimaana.

19th type mirror is described in Darpana prakarana:

Uraga-twak or snake-scales, pancha-mukha, vyaaghradanta or tiger's tooth, sand, salt, mercury, lead, white gum or shwetaniryaasa, mrittikaa, sphaatika or alum, ruruka, veera or red lead? mrinala or lotus tendril, ravikarpata, chanchola or red castor, vaalaja, panchapraanasaara or urinal salt of man, horse, ass, ox, and sheep, or ammonium chloride, shashodupa or benzoin shoot. These 18 ingredients in the proportion of 3, 7, 5, 22, 4, 15, 2, 5, 20, 7, 30, 15, 40, 23, 27, 13, 19, 18, purified thrice, filled in matsya moosha crucible, placed in nalikaa furnace, and with the aid of gomukhee bellows boiled to 99th degree, and poured into the mould, will yield pingala mirror.

Pattikaabhraka Yantra:

Says Kriyaasaara,

"In order to safeguard against the fires generated by the juxtaposition of planets during its course, the pattikaabhraka yantra is to be installed in the centre of the vimaana."

It is said in "Yantrasarvasva,"

In the course of its planetary motions, two planets sometimes get too near each other, when by the conflict of their giant forces, fires will burst out. They are known as jwaalaamukhee or flame-tongued, and will destroy the vimaana and those inside it. As protection against it, pattikaabhraka yantra should be installed in the vimaana.

The yantra should be made out of the 3rd type of mica amongst the 3rd group of its classification.

It is said in "Shownakeeya,"

"The names of mica belonging to the 3rd group are shaarada, pankila, soma, maarjaalika, rakta mukha, and vinaashaka. The yantra should be made of soma variety."

Somaabhra is described in Loha-tantra:

"It is sky-coloured, fine, strong, absorbent, cure for eye diseases. Its touch is cooling to the body. It has diamond content, and is cure for urinal trouble. It exhibits scarlet lines with whorls. These are the qualities of somaabhraka."

The mica should be purified twice with brinjal and mataa seed oil and melted, and a pattika two feet wide and of arm's length high should be fashioned. A koorma peetha or tortoise-shaped foot-plate 16 inches wide and arm's length high should be made with vaari vriksha. A shanku or pivot should be made like the pattika. Revolving wheels with keys and shoundeerya manis or crystals should be fixed. From the main, centre wires should be attached to it to the end of the pattika. On the other side an ivory vessel should be fixed, and filled with shyvaala acid, and adding mercury ravichumbaka manis or crystal should be placed in it. The wires should be connected to the inside of the vessel containing these things. From the pivot it should be covered with shringinee. The root of the naala or pivot should be fixed to face the sky. The mica shanku with five revolving wheel key attachments should be fixed in the centre of the peetha, and the acid purified pattika should be fixed on top of it in the centre of the vimaana covering.

When the jwaalaamukhee erupts from the planetary contact and reaches the direction of the vimaana, the main switch of the pivot should be operated, a cold wave will arise through the wires from the acid vessel, and passing through the five wheels reach the pattika, and contacting the jwaalaamukhee force will draw it and push it to the crystal in the centre of the enclosure, and the crystal will thrust it through the naala or pipe to the outside atmosphere where the flame force will get extinguished.

Soorya Shaktyapakarshana Yantra
or Solar heat extracting Yantra:

In order to relieve the excessive cold of the winter months, the soorya shaktyapakarshana yantra should be installed on the vimaana.

Says Yantra Sarvasva,

"In order to protect from the cold of the 4 winter months the solar heat storing machine is now explained. The 27th kind of mirror capable of capturing solar heat is to be used in its making."

It is said in Darpana prakarana:

Sphatika or alum, manjula or madder root, sea-foam, sarja salt or nation, sand, mercury, garada or aconite, kishora or wild

liquorice, gandhaka or sulphur brimstone, karbura or yellow orpiment, praanakshaara or ammonium chloride, in the proportion of 12, 1, 5, 1, 13, 12, 8, 10, 27, 4, 3, 7, 8, 5, 1. 5, 8, 3, 9, 2,
purified, to be filled in antarmukha crucible, placing it in shu-
ka-mukha furnace, and boded. Then pour it into antarmukha
yantra or vessel and turn the churning key. When cooled in the
mould a fine, light, strong, golden. coloured, solar heat collecting glass will be formed.

From this glass prepare a pattika or plank, 80 inches long,
20 inches wide, and 1 inch thick. Three spots are to be marked
on it. Two naalas or pipes, of arm's size, with 10 inch mouth,
crescent shaped peetha should be prepared. Another peetha, 2
feet long, and 6 feet high, should be prepared. The crescent
shaped peetha should be fixed in it. On its two sides the 2 naalaas should be fixed. Between them a pivot 88 inches long and 3
inches wide should be fixed. The other pattika should be fixed
on its top. At its 3 marked spots, lotus shapes with petals made
of the above glass with two faces with goblets on them should
be fixed. The two naalaas or pipes should be filled with shyvaala
or moss acid and shrini acid. Chhaayaamukha crystal should be
placed in them. At the foot of the shanku jyotsnaa acid should
be placed. Cold absorbing wires with key switches with ball
bearings should be fixed in the jyotsnaa acid. The wires should
be taken between the neighbouring naalas, taken round the two
lotus positions on the sides of the pattika, and then made to
surround the central lotus, and lead on and placed inside the
jyotsna acid. Then the other naala should be made to cover the
acid vessel, and fixed so as to have its opening through the bottom of the vimaana.

On the approach of winter cold into the vimaana, the main
wheel at the foot of the shanku should be turned at high speed.
That will energise the head wires of the pattika, making the lotus petals active, and the wind will draw the cold and pass to the
central acid vessel through the wires, and the acids in the 2 naalas
will draw in the cold and pass to the chaayaamukha mani, which
by its own force will pass the cold force to the jyotsnaadravaka,
which will eject it through the naala to the atmosphere outside

for being dissolved. The vimaana will thus be saved from the cold force through the soorya-shaktyapakarshana yantra.

Apasmaara dhooma prasaarana yantra
    or poison-gas fume spreading machine:

Says Kriyaasaara,

"When the enemy plane is trying to destroy your vimaana, Apasmaara dhooma prasaarana yantra should be provided in the vimaana to combat it."

Yantra Sarvasva Says:

"Apasmaara dhooma prasaarana yantra is prescribed for protection of vimaana from enemy planes. It should be manufactured with khoundeera metal only, and not with anything else."

Kshoundeera loha is described in Lohatantra:

8 parts of kshwinka or zinc, 5 parts of mercury, 7 parts of krowncha alloy, 3 parts of kaanta or steel, 4 parts of hamsa or metallic silver, 1 part of maadhweeka, and 5 parts of ruru, these ingredients to be purified and filled in crucible, placed in chhatreemukha furnace, and with the acid of surasa bellows heated to 100 degrees heat, and cast in mould will yield excellent kshoundeera alloy.

Filling this loha in pattikaayantra, applying 300 degree heat and churning a fine strong pattikaa will be formed. With that, a shape like bellows, 5 arm's length high, and 3 arm's wide should be formed. It should be provided

p. 81

with a mukha-naala or nozzle 6 feet in size. Its mouth should be like that of peshanee yantra. The opening should be covered and keyed. Three satchels should be attached at its bottom. In the middle an aavarana or covering with hare-skin, circular and provided with switch. Smoke or gas filling switch should be provided at its base. Above it a choorna paatra or powder vessel should be fixed. The switch key should be beneath the middle of the vessel. Thus four bellows should be prepared.

When the enemy plane's attack is expected, the aavaaraka bhastrika or enveloping bellows should shield the vimaana, and the 4 bellows should be fixed on the dikpeethaas or side seats above the aavarana or covering, and electricity should be applied to the choorna-paatra or powder keg. Immediately the powder becomes smoke. The bellows' mouth should be opened and the key turned. The emerging smoke fumes will enter the 4 small bellows, and from them reach the central kunda and spread all around and reach the bellows' mouth. Then on turning that key, the fumes will be emitted from all the bhastrikaas or bellows, and encompass the enemy plane and disable its occupants. That plane will be destroyed and the danger to one's vimaana overcome.

Stambhana Yantra
or Halting machine:

Kriyaasaara says,

When power is generated by conflicting forces in the water-charged regions, shrieking hurricanes and whirl-winds will arise and set out in a mad career of destruction. To safeguard against their onslaughts, the Stambhana yantra should be installed in the bottom of the vimaana.

Yantra sarvasva says,

In order to protect the plane from the attacks of giant wind blasts, vimaana stambhana yantra is described as follows:

A peetha, one fourth the size of the vimaana floor, quadrangular, should be made out of vakratunda metal, three feet in thickness. In its 8 quarters spots should be marked for fixtures. An enclosure with openings, revolving toothed wheels made of the same metal, wheel rods with revolving keys, a metal band

which is to encircle it thrice, toothed wheels, pivots, and switches and hinges, and three- stranded wire ropes, should also be of
the same metal. In the 8 selected spots naalaas or pivots with
wheels and wiring should be fixed. Through the naala or pipe at
the contraction switch of the vimaana, wires should be passed
through the other naalas to the central pivot, and tied at the
foot of the revolving wheel. When the dreaded wind current
is observed the switch or wheel for the contraction or folding
of the expanded vimaana parts should be turned, as also the 8
side wheel turning switches. That will reduce the speed of the
vimaana. Then the switches of the 8 pivots on the peetha or
foot-plate should be turned. The entire speed of the vimaana
will be extinguished thereby. The wheel at the central pivot of
the peetha should then be turned, so that the vimaana will be
halted completely. Then the switch of the plane-wings should
be turned. The flapping of the wings will produce winds which
will encircle the vimaana and form a globe protecting it. Then
by turning the switch of the brake-rod, the vimaana becomes
motionless. Therefore the Yaana-sthambhana yantra should be
fixed at the bottom centre of the vimaana.

Vyshvaanara-naala yantra:

Kriyaasaara says,

For the purpose of providing fire for passengers to perform
agnihotra or daily fire rituals, and for the purpose of cooking
food, Vyshwaanara naala yantra is to be fitted up at the navel
centre of the vimaana.

Yantra-sarvasva says,

To provide fire for travellers in vimaanas, vyshwaanara naala
yantra is now described. A 2 feet long and 12 inches wide peetha
or foot-plate should be made out of naaga metal, quadrangular
or circular in shape. Three kendras or spots should be marked
thereon. Three vessels should be made of copper and karpara
or (black jack?) zinc blended metal. One vessel should be filled
with sulphur- brimstone acid. Another should be filled with

rookshaka bdellium acid, or croton seed acid? And manjishta or madder root acid should be filled in another vessel. The three vessels should be placed on the 3 kendra spots on the peetha.

In the sulphur acid vessel the prajwaalaka mani or flame producing crystal should be placed. In the rookshaka acid vessel the dhoomaasya mani or smoke crystal should be placed. In the manjishtha acid vessel the mahoshnika mani, or heat producing crystal should be placed.

In the places in the vimaana where kitchens are located, and where sacred agnihotra fires are needed by passengers, keelaka sthambhas or pivots should be fixed. The acid vessels should be connected with power wires from the central pivot. The wires should be attached to the manis or crystals in the acid vessels. At the top of the central pole jwaalaamukhee mani should be fixed in the centre of chumbakee keela with glass covering. On either side of it sinjeeraka mani and dridhikaa mani should be fixed. From each mani a wire should be stretched from the top of the central pole and fixed at the granthikeela at the foot of the pole. From there up to the cooking spots and agnihotra spots, a circle should be formed like a kulya, and metal tubes should be fixed therein. Wires should be drawn through the tubes to the fire places and fixed to the kharpara metallic pattikas therein.

First the bhadramushti keelaka should be revolved. The acid in the vessel will become heated. The heat generated in the rookshna acid will pass into the manjishtha mani, and generate smoke in the mahoshnika mani. By the force of that acid intense heat will be generated. And by the heat generated in the sulphuric acid vessel flames will erupt in the prajwaalika mani. The smoke, heat and flames will pass through the wires to the sinjeeraka, dridhikaa and jwaalaa-mukhee manis. Then the chumbaka wheel should be turned vigorously, whereupon the smoke, heat, and flames will reach the key at the top of the central pole. And on that keelee being turned, they will reach the central switchboard keelee at the foot of the pole. When that switch is put on, the heat and flames will reach the metal bands of the cooking ovens and religious fire places, and generate fires. Therefore vyshwaanara naala yantra should be fixed at the navel centre of vimaana.

We have so far dealt with anga yantras or constituent machines of the vimaana. We proceed next to deal with Vyoma-yaanas or Aeroplanes.

SIXTH CHAPTER

Atha Jaatyadhikaranam: Varieties of Vimaanas.

Maharshi Bharadwaaja:

☞ **Jaati tryvidhyam Yugabhedaad Vimaanaanaam. Sootra 1.**
"Three types according to changing Yugas."

Bodhaananda Vritti:

According to the differences in yugas, there are three different types of planes:

Having dealt with the constituent mechanical parts of the Vimaana, we shall now deal with the Vimaanas according to their different classes.

The sootra indicates that there are different types of planes, and that they are of 3 types.

In the Krita Yuga, Dharma or Righteousness was four-footed, that is, it was four-square, fully established, all paramount, and it was adhered to implicitly by men. The men were inherently noble-born and were possessed of remarkable powers. Without needing to go through yogic discipline to attain special powers, or practise mantras which secured extraordinary results, the men of that yuga, merely by their devotion to dharma, became Siddhapurushaas or gifted with superhuman powers. They were virtuous men and men of learning and wisdom. Going in the sky with the speed of wind by their own volition was natural to them. The eight super-sensory, and now superhuman, attainments, known as animaa, mahimaa, garimaa, laghimaa, praaptih, praakaamya, eeshatwa, and vashitwa, were all possessed by them. That is, animaa is assumption of infinitesimal shape;

mahimaa is growing into gigantic shape; garimaa is becoming astonishingly heavy; laghimaa is becoming weightless; praaptih is securing any desired thing; praakaamyam is becoming rid of desires; eeshatwa is attaining paramountcy; and vashitwa is becoming extremely pliant.

*   *   *   *

I. Krita yuga—1.728.000 years.
II. Threthaa yuga—1.296.000 years. .
III. Dwaapara yuga—864.000 years
Kaliyuga—432.000 years.

Therefore in Krita Yuga, or first epoch, the ancients say, there were none of the three classes of Vimaanas.

Krita Yuga passed; and Tretaa Yuga commenced. Dharma then became limp of one foot. It served with 3 feet only, and grew gradually less efficient. So men's minds became dense, and the conception of Vedic truths, and anima and other super-sensory powers, became scarcer. Therefore, by the corrosion of Dharma or righteousness, men lost the power of flying in the sky with the speed of wind.

Perceiving this, God Mahadeva, desiring to confer the power of understanding the Vedas properly on the Dwijas, or brahmins, kshatriyas, and vysyas, graciously descended on earth in the form of Dakshinaamurthy, and through the instrumentality of Sanaka and other anointed sages, classified the Veda mantras, and then bestowing his benedictory glance on the worshipping Munis or ascetics, he blessed them with the gift of Vedic perception. And then to ensure that they were properly receptive, he embraced them and entered their hearts and illuminated their memories. The munis, overwhelmed by the Divine grace, aglow with horripilation, with voice choked with emotion, praised the Supreme with shata-rudreeya and other hymns, and manifested profound devotion.

Pleased with their receptiveness, divine Dakshinaamurthy, favouring them with a benign glance, and with smile on his face, said to them, "Till now you have been known as "Munis" or ascetics. Henceforth, having by my grace attained insight into the Vedas, you shall be known as "Rishis" or seers. You will cultivate the Vedic mantras, and practising HYPERLINK http://www.bibliotecapleyades.net/vimanas/vs/errata.htm#17 celibacy, you will adore the divine Goddess of the Vedas, and winning her favour, and approaching the Great God Easwara by Yogic Samadhi, ascertain His mind, and by His and my grace, rising to the pinnacle of intellectual vision, become adepts in the meaning and purport of the Vedas; and confirming by them your own experiences and meditative introspection, you will create the Dharmashaastras or moral codes, Puranas and Itihaasas, and physical and material sciences, in conformity to the truths of the Vedas, for the benefit of mankind. And for travelling in the sky, propagate the art of manufacturing Vimaanas, and for attaining wind-speed, evolve Ghutica and Paadukaa methods through Kalpashaastras or scientific treatises."

Then those munis or seers, enshrining in their hearts God Mahadeva in the form of Dakshinaamurthy, produced the Dharmashaastras or ethical codes, epics, chronicles, manuals on rituals, treatises on the arts and sciences, ritualistic and sacrificial codes, in conformity to the Vedas, and propagated them among men. Amongst them it is said that there are six treatises bearing on the manufacture of Vimaanas produced by the ancient seers. In them are described three classes of vimaanas, known as maantrikaas, taantrikaas, and kritakaas, capable of flying everywhere.

It is said in Vimaana Chandrika,

"I shall indicate the different kinds of vimaanas. In Tretaa yuga as men were adepts in mantras or potent hymns, the vimaanas used to be produced by means of maantric knowledge. In Dwaapara yuga as men had developed considerable tantric knowledge, vimaanas were manufactured by means of tantric knowledge. As, both mantra and tantra are deficient in Kaliyuga, the vimaanas are known as kritaka or artificial. Thus, owing to

changes in dharma during the yugas, the ancient seers have classified the vimaanas of the 3 yugas as of 3 different types."

"Vyomayaana Tantra" also says,

"By the influence of mantras in Tretaa, vimaanas are of
maantrika type. Owing to the prevalence of tantras in Dwaapara, the vimaanas are of taantrika type. Owing to decadence of
both mantra and tantra in Kaliyuga, the vimaanas are of artificial type." Thus 3 classes of vimaanas are mentioned in shaastras by ancient seers.

In "Yantra Kalpa" also,

"Vimaanas are classified into mantra and other varieties by
experts according to differences in yugas. They are defined as
maantrika, taantrika, and kritaka."

The same is expressed in "Kheta yaana pradeepika," and also
"Vyoma Yaana Arkaprakaashikaa."

Thus according to shaastras vimaanas are divided into 3
classes, on the basis of differences in the modes of their manufacture.

Maharshi Bharadwaaja:

☞ **"Pancha-vimshan Maantrikaaha Pushpakaadi
Prabhedena" Sootra 2.**
"Maantrika Vimaanas are of Pushpaka and other 25 Varieties."

Bodhaananda Vritti:

In the previous sootra vimaanas were specified as of 3 types
owing to differences in the 3 yugas. In this sootra maantrika vimaanas or vimaanas flying by maantrik power are said to be 25.

Shounaka Sootra says,

Maantrika vimaanas in Tretaayuga are 25. Their names are pushpaka, ajamukha, bhraajasvat, jyotirmukha, kowshika, bheeshma, shesha, vajraanga, dyvata, ujvala, kolaahala, archisha, bhooshnu, somaanka, panchavarna, shanmukha, panchabaana, mayoora, shankara, tripura, vasuhaara, panchaanana, ambareesha, trinetra and bherunda.

In Maanibhadrakaarikaa,

The vimaanas of Tretaayuga are 32 of the maantrika type. Their names as given by Maharshi Gowtama are Pushpaka, ajamukha, bhraaja, swayamjyoti, kowshika, bheeshmaka, shesha, vajraanga, dyvata, ujvala, kolaahala, archisha, bhooshnu, somaanka, varnapanchaka, shanmukha, panchabaana, mayoora, shankara priya, tripura, vasuhaara, panchaanana, ambareesha, trinetra, and bherunda, etc.

Maharshi Bharadwaaja:

☞ **"Bhyravaadi Bhedaat Tantrikaa-shshat-panchaashat."**
Sootra 3.
"Taantrika Vimaanas are of Bhyrava and other 56 varieties."

Bodhaananda Vritti:

In previous sootra the names of maantrika vimaanas were mentioned: In this sootra the names of taantrika vimaanas of Dwaapara yuga are mentioned.

In shape, movement and speed there is no difference between maantrika and taantrika vimaanas. There is however one difference in taantrika vimaanas, that is, the way in which the shakti or power at the junction of sky and earth is incorporated.

Lalla also says,

There is only one difference between taantrika vimaanas and maantrika vimaanas: the adaptation of the power of sky and

earth. In shape, and movement. variations, they are identical.
The taantrika vimaanas are of 56 varieties.

In Shounaka Sootra,

In Dwaapara taantrika vimaanas are 56. Their names are,
bhyrava, nandaka, vatuka, virinchi, vynateya, bherunda, ma-
karadwaja, shringaataka, ambareesha, sheshaasya, saimtuka,
maatrika, bhraaja, paingala, tittibha, pramatha, bhoorshni,
champaka, drownika, rukmapunkha, bhraamani, kakubha, kaalabhyrava, jambuka, garudaasya, gajaasya, vasudeva, shoorasena,
veerabaahu, bhusunda, gandaka, shukatunda, kumuda, krownchika, ajagara, panchadala, chumbuka, dundubhi, ambaraasya,
maayooraka, bheerunalika, kaamapaala, gandarksha, paariyaatra,
shakunta, ravimandana, vyaaghra, mukha, vishnuratha, sowarni-
ka, mruda, dambholi, brihathkunja, mahaanata, etc.

In Maanibhadrakaarikaa:—

In Dwaapara yuga taantrika vimaanas are said to be 56.
Their names according to sage Gowtama, are bhyrava, nandaka,
vatuka, virinchika, tumbara, vynateya, bherunda, makaradhwa-
ja, shringaataka, ambareesha, sheshaasya, symhika, maatruka,
bhraajaka, pyngala, tittibha, pramatha, bhoorshnika, champaka,
drownika, rukmapunkha, bhraamanika, kakubha, kaalabhyirava,
jambuka, gireesha, garudaasya, gajaasya, vasudeva, shoorasena, veerabaahu, bhusundaka, gandaka, shukatunda, kumuda,
krownchika, ajagara, panchadala, chumbaka, dundubhi, ambaraasya, mayoora, bheeru, nalikaa, kaamapaala, gandarksha, paariyaatra, shakuntaka, ravimandana, vyaaghramukha, vishnu ratha,
souvarnika, mruda, dambholee, bruhatkunja, mahaanata.

These 56 are taantrika vimaanas of Dwaaparayuga.

Maharshi Bharadwaaja:

☞ **"Shakunaadyaah Panchavimshat Kritakaah." Sootra 4.**
"Shakuna and other 25 types of Vimaanas are Kritakaah."

Bodhaananda Vritti:

In shape and movements there is no difference in the vimaanas, except in the matter of the use of mantraas and tantraas. The kritaka or artificial vimaanas are of 25 varieties.

According to Shownaka sootra

"Tishyay kritaka bhedaah panchavigamshatih! teshaam naamaanyanukramishyaamah: shakuna sundararukma mandala vakratunda bhadraka ruchaka vyraaja bhaaskara gajaavarta powshkala virinchi nandaka kumuda mandara hamsa shukaasya somaka krownchaka padmaka symhika panchabaana owryaayana pushkara kodandaa iti."

Says "Maanibhadra Kaarikaa":

In Kaliyuga, the kritaka or artificial vimaanas are said to be 25. Their names are given below as indicated by sage Gowtama: shakuna, sundara, rukmaka, mandala, vakratunda, bhadraka, ruchaka, viraajaka, bhaaskara, gajaavarta, powshkala, viranchika, nandaka, kumuda, mandara, hamsa, shukaasya, sowmyaka, krownchaka, padmaka, symhika, panchabaana, owryaayana, pushkara, and kodanda.

Maharshi Bharadwaaja:

☞ **"Raaja-lohaadeteshaam Aakaara Rachanaa." Sootra 5.**
"These should be built out of Raajaloha."

Bodhaananda Vritti:

These 25 kinds of vimaanas are to be made of Raajaloha metal only.

Says Kriyaasaara,

In manufacturing artificial aeroplanes the best of metals are those known as Ooshmapaa or heat-imbibing or heat resisting metals. Out of them the variety known as Raajaloha or king of metals is most suited to Shakuna and other vimaanas.

Three kinds of metals, soma, soundaala, and maardweeka, in the proportion of 3, 8, and 2, adding borax, to be filled in crucible or smelter, and placed in furnace, and heated to 272 degrees, and melted thoroughly, and churned, will result in the alloy Raajaloha.

Vishwambhara also says,

"In the science of metals, for the manufacture of aeroplanes, 16 types of Ooshmapaa or heat-sucking lohas or metals are the very best. The fourth in that series, is called Raajaloha. Out of that alone should shakuna vimaana be constructed."

The parts of shakuna vimaana are:

Peetha or floor board; hollow mast; three wheeled keelakas with holes; 4 heaters, air-suction pipes, water jacket, oil tank, air heater, chhullee or heater, steam boiler, vidyud-yantra or electric generator, air propelling yantra, vaatapaa yantra or air-suction pipe, dikpradarsha dhwaja or direction indicating banner, shakuna yantra, two wings, tail portion for helping vimaana to rise, owshmyaka yantra or engine, kiranaakarshana math or sun-ray attracting bead. These 28 are parts of Shakuna vimaana.

The construction of the vimaana:

The floor-board or base should be made of levelled Rajaloha sheet, shaped quadrangular, circular, or cradle shaped. The weight of the peetha should be one-hundredth of that of the plane, and its width should be half the height of the vimaana. In the centre of the peetha the hollow mast should be fixed with screw joints.

Lalla defines the mast in "Yantra kalpataru". The stambha or mast should be made of haatakaasya metal and not otherwise.

Haatakaasya metal is described in "Lohatantra": 8 parts of suvarchala or natron, 16 parts of laghu-kshwinka or light zinc, 18 parts of lagbu bambhaari, and 100 parts of copper, filled in smelter, placed in koorma vyaasatika furnace, and with the aid of mahormi bellows boiled to 307 degrees, will yield haatakaasya metal.

The Peetha

The height of the peetha should be 80 feet. It should be 56 feet in length and breadth, 70 feet high on the north and south sides. The tip should be three-cornered. This is for shakuna vimaana.

Naalastambha or Hollow Mast:

At the bottom the mast should be of 35 feet diameter outside, and 30 feet inside. At the middle the mast should be of 25 feet diameter outside and 20 feet inside. Higher up it should be of 20 feet diameter outside and 15 feet diameter inside; The height of the mast should be 80 feet. It should be made of Raajaloha. In order to fix the mast in the peetha screw joint should be made. And in order to adjust the air-speed as required, 6 wheels should be inserted inside the mast.

The Wheels:

Inside the mast at the height of 4 feet above the peetha, three wheels should be provided, of 15½ feet diameter, with holes. The wheels above and below should be fixed with bolts, and unmoving. In order to revolve the middle wheel keys should be fixed outside on the mast, As there are holes in the wheels, as two wheels do not move, and as the middle wheel revolves in a group with the other two wheels, movement of air is, allowed or stopped by the turning of the key outside.

Similarly at the height of 44 feet above the peetha three wheels corresponding to those below should be fixed and operated similarly.

Window dome:

The window dome should be of 15½ feet outside circumference. Its inside should be five feet wide and it should be 2 feet high. It should be fixed on the top of the mast.

Sun-crystal:

A sun crystal 7 feet round, and 2 feet wide and 2 feet in height should be fixed so as to crown the window dome.

10 feet above the bottom peetha, on a floor-board 3 inches thick, three floors or tiers should be built, each 14 feet high, with 3 inch floor-boards, the upper two floors being supported by pillars fixed at 10 feet intervals with screw joints and strong bolts. In the four corners 4 heating yantras should be fixed, 10 feet in circumference and 8 feet high. On the ground floor along the supporting pillars accommodation for passengers should be provided in the form of individual boxes.

On the second floor booths should be constructed to accommodate the anga-yantras, or the various mechanisms recommended for the safety of the vimaana. It should be 60 feet wide and 14 feet high with 3 inch thick ceiling board.

The third floor should be 40 feet wide and 14 feet high.

The partitioning boxes for passengers as well as the booths of the various machines should be divided off by railings starting from the hollow mast to the side walls in all the four directions.

Beneath the ground-floor board a 7 feet high cellar should be constructed. In it the several necessary yantras should be located. In the centre is the foot of the hollow mast. On the four directions from it 4 air pumping machines should be fixed. In order to stimulate them 4 steam engines also should be installed. On the two sides of the vimaana two air expelling machines, and an air heater machine, and 2 machines to keep the heater supplied with air from outside, should be erected.

In order to enable the wings on either side to spread and flap, proper hinges and keys should be provided for, safely fixing them to the sides of the vimaana, and for enabling them to fold and open easily.

The revolving tractor blades in the front should be duly fixed to the heating engine with rods so that they could dispel the wind in front and facilitate the passage of the vimaana.

The wings are two, one on each side, very strongly fixed to the vimaana with bolts and hinges. Each wing should be fixed in a 11 foot scabbard up to 20 feet length, where it would be 10 feet wide, widening further up to 40 feet at the end of its 60 feet length, besides its. first 20 feet of scabbard length.

The tail should be 20 feet long, and 3½ feet wide at the start, and 20 feet wide at the end.

The air-blower and heater:

The length of the air-blower should be 15 feet, and width 3 feet. The naalaas or pipes should be 3 feet wide, and their outer circumference should be 4 feet. The rods and hinges and other equipment should be suitably prepared.

The vaatapaa yantra or air blower should be 12 feet long and 9½ feet wide. Inside it should be covered with circling wires. A pipe should be fixed inside, for air flow. By the hot oil fumes from the heated tank, the air becomes heated and should be passed into the owshmya yantra or heater, while the cold air from outside should also be let in. Tubes and fixings should be provided in the yantra. In order to emit the fumes of the oil flames to the outside, a 6 inches pipe should be fixed from the yantra to the foot of the mast. Air blowers should be installed with 10 feet wheels to pump in fresh cold air from outside.

To the east of the air machine should be placed a light burner in order to aid combustion of the oil. An electric generator provided with switches should light the burner. When the light is off the oil should be kept duly sealed. A rope should be tied to the tail joint, for the pilot to manipulate the fluttering of the tail to help the ascent or descent of the vimaana. Similarly ropes should be tied to the hinges of the two wings, and passed to the pilot like reins, so that he might spread them out or close them as needed.

Ten feet beneath the passenger floor of the vimaana, to a height of 2½ feet from the bottom plate there should be a cel-lar-like enclosure. The bottom of the vaatanaala mast should be fixed in its centre with firm screw joints. In this cellar should be located two oil tanks 15 feet by 9½ feet by 4 feet, with water jackets.

Four bellows of 15 feet by 2½ feet by 6½ feet, should be provided for storing the air pumped in by the air-blowers, and letting it out as required.

And underneath, on all the four sides wheels of 7 feet cir-cumference should be fixed for the movement of the vimaana on the ground.

This vimaana is named **SHAKUNA VIMAANA**.

SUNDARA VIMAANA

Maharshi Bharadwaaja:

☞ **"Sundarothha." Sootra 6.**
"Next Sundara."

Bodhaananda Vritti:

Next Sundara vimaana will be described. It has got 8 constituent parts.

First peetha or ground plate, smoke chimney, 5 gas-engines, bhujya metal pipe, wind blower, electricity generator, and four-faced heater, and vimaana nirnaya, or outer cover.

The Peetha or ground plate:

It should be made of Raajaloha metal only. It should be square or round, and of 100 feet in circumference, or any other desired size. It should be 8 feet thick. Seven times the peetha has to be heated with manchuka or madder root oil. Then spots should be marked in it at 10 feet distance from each other, totalling 24. The size of each kendra or centre is 15 feet. In the centre a dhooma-prasaarana or fume distributing naala or pipe 12 feet high should be erected.

Naalastambha, hollow mast:

The naalastambha should be 56 feet high, and 4 feet in diametre. For storing gas, at its base, a 8 feet long, circular, and 4 feet high vessel should be provided. A six feet size water vessel should be arranged. A 4 feet size oil tank should be fixed at its centre. At its foot an electric storing crystal of 1 foot size should be fixed with necessary hinges and keys.

The vessel should be filled with 12 parts of dhoomanjana oil, and 20 parts of shukatundika or bignonia Indica? (egg-plant?) oil, and 9 parts of kulakee or red-arsenic oil. To conduct electricity, two wires should be passed through the pipe and fixed to the crystal. In the middle of the naalastambha or mast, for the smoke fumes to be restrained or speeded out, triple wheels with holes should be fixed. In order to work the wheels from outside, two right turning and left turning wheels should be attached outside the pole, and connected to the wheels inside. Three wires should be drawn inside the naala and fixed at the foot, the middle, and at the top.

Dhoomodgama Yantra:

Because it ejects smoke fumes with speed it is called Dhoomodgama yantra.

Hima samvardhaka, soma, and sundaala, in the proportion of 32, 25, and 38, should be filled in pipe crucible, placed in chakra-mukha furnace, and with the help of ajaamukha bellows heated to 712 degrees and properly churned. It will yield excellent dhooma-garbha alloy. With that alloy the dhoomodgama yantra should be constructed.

Underneath the centre of the 15 feet long peetha, for the control of the gas fumes a 10 feet high pipe with right revolving wheel should be fixed. On its 2 sides, to south and north, 2 water steam pipes should be erected. At the foot of the 2 pipes 4 feet long 3 feet high pots should be formed for containing the fumes. Two pipes shaped like goblets, 1 foot by 8 feet by 3 feet, should be fixed at the top of the fume container. A water vessel at its foot, and an oil-vessel at its centre, and in front of it the switches of the electric ray crystals, as in the dhooma prasaarana naala stambha.

On either side of the heat tube, two water jackets should be placed. A pipe with wires should be taken from the electric generator and connected to the hinges of the crystals. Electric current of 80 linkas should be passed to the crystals, whose motion will cause friction and generate heat of 100 degrees (kakshyas). Thereby the oil in the vessel will get heated and boil and emit fumes.

The electric power should then be passed through the smoke pipe between the two water jackets. By this the water will be converted into hot steam. The oil fumes should be filled in the oil fume pipe and the steam in the steam pipe. Then by operating the switches, both the fumes will fly up at 500 degree temperature. The switches should restrain the fumes or pump them out as needed. 40 such yantras should be prepared and should be fixed on the peetha in groups on the four sides. Then connected with the bases of the dhooma-naalas, sundaalas or elephant trunks, one foot wide and 12 feet high should be erected on the four sides, to enable the vimaana to fly with speed.

Sundaala is described by Lallachaarya:

The sundaala should be installed. for using the oil fumes and steam fumes for the motion of the vimaana. There are. varieties of ksheera vrikshaas or milk-trees according to shaastraas. Vata or banyans, manjoosha or madder root, maatanga or citron?, panchashaakhee (five branched), shikhaavalee (crested), taamra sheershnee (copper-crested), brihatkumbhee (big bellied), mahishee, ksheeravallaree, shona parnee (crimson-leaved), vajramukhee, and ksheerinee (milky). From these the ooze or milk should be collected, and in the proportion of 3, 5, 7, 10, 11, 8, 7, 4, 7, 30, 12, filled in a vessel. Then granthi metal, naaga or lead, vajra, bambhaarika, vynateya, kanduru, kudapa, and kundalotpala, these in equal parts should be filled in the vessel in equal proportion to the milk contents, and boiled with 92 degree heat. Then the molten liquid should be filled in the milk-cloth machine, and churned. When cooled and put through the levelling machine, it will yield a strong, soft, cool, heat proof, and uncuttable ash-coloured cloth sheet.

This cloth should be boiled in rouhinee taila or oil of black hellebore for 3 yaamaas or 9 hours, and then washed with water. Then it should be boiled in atasee or linseed oil as before.

Then it should be kept in ajaa-mootra or goat's urine for one day and kept in the sun. Then it should be dried and painted with kanakaanjana paint and dried, Then the cloth will glow with a golden hue. With this cloth should be made the shundaala or elephant trunk, 12 feet high, 1 foot round, and with pipe-like opening inside.

Two mechanisms for rolling it and unrolling it should be properly attached to it. By the rolling switch the shundaala will coil round like a snake and remain on the floor. By the unrolling switch it will uncoil and stand erect like a raised arm. From the fume generating yantra connecting links to the shundaala should be provided for the fumes to pass through it to the outside air. And to attract outside air into the sundaala a pump-like arrangement should be provided as in an inflator.

Three switches should be provided as in the water tapping yantra. By revolution of its wheel the fumes will go out through the shundaala and 82 linka of fresh air will come in. The direction in which the fumes will emerge from the shundaala will be the direction of the course of the vimaana. The 3 wheels in the shundaala will cause the vimaana to wheel around or make ascent, or to drop height.

At the foot of each dhoomodgama yantra 2 shundaalas should be duly fixed. And on the 4 sides of the dhoomaprasaa-rana-naala- stambha 4 shundaalas should be erected.

In order to protect against the intense heat from fire and sun inside and outside the vimaana, it should be provided a covering made of the 6th type of Ooshmapaa loha or heat-proof metal. At the top and bottom and on the sides keys should be provided for the movement of the fumes. 40 such dhoomodgama yantras should be properly fixed in the selected spots of the peetha with screw fittings. The vimaana will be enabled to fly smoothly by so doing.

ELECTRIC DYNAMO

Says Yantra Sarvasva:

There are 32 kinds of yantras for generating electricity, such as by friction, by heating, by waterfall, by combination, by solar rays, etc. Out of these, saamyojaka or production by combination is the one most suitable for vimaanas. Its manufacture is explained by Sage Agastya in Shaktitantra:

The peetha or foot-plate should be made of saamyojaka metal, 35 feet in diametre. 5 spots should be marked in it in a circle, 5 feet in diametre, with a spot in the centre. Vessels should be prepared for each kendra, 4 feet wide, 2 feet high, shaped like a pot. On each a cylindrical pipe 1 foot wide and 1 foot high, should be fixed. The top of the cylinder should be 4 feet wide and round.

Then get a Jyotirmukha or flame-faced lion's skin, duly cleaned, add salt, and placing in the vessel containing spike-grass acid, boil for 5 yaamas or 15 hours. Then wash it with cold water. Then take oils from the seeds of jyothirmukhee, or staff-tree, momordica charantia, and pot herb, in the proportion of 3, 7, and 16, and mix them in a vessel, add 1/64 part of salt. The skin should be immersed in this oil and kept for 24 days in solar heat. It will get a scarlet sheen. The skin should be cut to the size of the top opening of the vessel cylinder, with 5 openings in it. Cover the cylinder with the skin with bolts. All the 5 vessels should be similarly covered, and placed in the 5 selected centres on the peetha. Then 16 drona measures of asses' urine, 16 linka measures of mined charcoal, 3 linkas of salt, 2 linkas of snake-poison, and 2 linkas of copper, should be filled in the vessel on the eastern side.

Then in the vessel on the western side, 7 vidyudgama mani or load-stone, 13 praana-kshaara or ammonium chloride, 22 hare-dung, should be filled. and made into a decoction. Two parts of camel urine should be mixed with one part of the above. Then 50 linkas of rhinoceros bones, 30 linkas of sulphur, and

16 linkas of tamarind tree salt, and 28 linkas of steel should be added to that. And 117 tatin-mitra manis should be placed in the centre of the vessel.

Next the following materials should be filled in the northern vessel:

Eleven parts of oil of apaamaarga or achyranthus aspera seeds, 32 parts of oil seeds of sarpaasya or mesua ferrea, 40 parts of ayaskaantha or oil of steel, in 83 parts of elephant's urine, all these to be put in the northern vessel and mixed together properly. Then add mercury, symhika salt, and paarvanika or bamboo rice, 30, 20, and 25 palas respectively, or 120, 80, and 100 tolas. Sun-crystal of the 800th type, mentioned in Maniprakarana, cleaned in oil, should be put in the vessel.

Next in the vessel on the southern side, put in grandhika draavaka or long-pepper decoction, panchamukhee draavaka, and shveta- punja or white liquorice decoctions, in proportion of 12, 21, and 16, and mix together, add cows' urine 5 parts more than the above liquids, 47 parts of jyotirmayookha root, 28 linkas of kaanta metal, 28th and 10th kind of kudupa 32 parts. 92 jyotirmanis purified in milk should be placed in it, according to Chaakraayani. This is the southern vessel.

Then in the central vessel electric current should be stored. That vessel should be made of chapala-graahaka metal only.

Chapala-graahaka metal is explained in Lohatantra:

Quick-lime, marble stone, lac, sowraashtra earth, glass, root of the elephant trunk tree, bark of karkata tree, cowries, cubeb pepper gum, in the proportion of 8, 11, 7, 27, 8, 5, 3, 7, and 12 parts of tankana or borax, to be filled in urana crucible, placed in kundodara furnace, and with 3 faced bellows, boiled to 427 degrees, will yield, when poured into the cooler and cooled, chapalagraahaka metal.

The electricity storage vessel should be manufactured as follows: A foot-plate 5 feet long, 8 feet high, 1 foot thick, half-moon shaped, should be made of above metal. The vessel should be shaped like a big pot, with a cylindrical top. It should have a glass covering. 2 pipes 3 feet wide 6 feet high should be fixed in the vessel in the northern and southern sides. They should also be covered with glass. Between the two pipes two wheels with hinges and switches etc. should be fixed. When the switches are put on or turned, causing the two wheels to revolve, electricity will flow from the bottom of the 4 vessels into the two pipes and ascend. Two tubes, 6 inches long, should be prepared, wound round with deer skin, tied with silk thread or silk cloth. The Vajramukhee copper wires cleaned with acids, should be passed through each tube, and taken to the two pipes in the vessel and be fixed with glass cups. 8 palas or 32 tolas of mercury should be placed in the energy container vessel. 391st vidyunmukha mani, wound round with copper wiring with mixing switch, should also be inserted. Then taking the wires in the pipes they should be connected with the wiring of the mani through the kaachakanku hole. In each of the vessels, excepting the middle one, two churning rods should be fixed in the centre. The rods should be made of steel or shakti skandha. They should be 3 feet high and 1 foot thick. Keys should be fixed in them for obverse and reverse churning. To the east of the churning machine wheels should be fixed for raising and lowering. An 8 inches high naala or tube should be fixed. On either side of it should be fixed 5 wheels of 5 inches height, like the wheel of the water lifting machine. 2 inches wide flat pattis made of shakti skandha metal should be passed from the wheels inside the Aavritta-naala to the keys of the wheels in the churning yantra. Then revolving wheels should be attached to the naalas or tubes of the stambha or big pipe. By the turning of these keys, it will operate like the turning of the churning rod back and forth as in churning curds by drawing and relaxing the rope ends.

Then according to Darpana-shaastra, four vessels, shaped like the bamboo cylinder used on the pounding mortar, should he made out of ghrinyaakarshana glass or solar-heat absorbing glass and fixed on the mouth of the 4 vessels.

The vessel is described by Lallaacharya: 8 inches wide and 1 foot high, and then 2 feet wide and 6 feet high, and at the top a 6 feet wide mouth.

25 palas or 100 tolas of bamboo salt, should be put in it. Then amsupaa mani or solar-ray crystal of the 325th kind, duly cleaned in acid, should be put in it with rice salt. Then rice hay should be spread over it tightly, and facing the sun. The rays from all sides are imbibed by them, and will enter the vessel daily to 105 degrees' strength. If kept thus in the sun for 12 days, 1080 linkas of electric power will be accumulated in each vessel.

In order to store this power in the storage vessel six inches long steel tubes should connect the bottom of the vessel with the storage vessel. They should be covered by deer skin and wound round with silk cloth or yarn. Two copper wires should be passed through the tubes and connected to the storage vessel. 100 palas of mercury should be put in the vessel. And a 391th type of sun crystal duly wired should be placed in the mercury, and the wires coming from the tubes should be connected to it.

The well-oiled keys in the 4 vessels should be revolved with speed, to 200 degrees heat, when the liquids in the vessels will be boiled by the heat rays. Then the keys should be hastened up to 2000 degrees. By the liquids in each vessel 800 linkas of electricity will be generated. The power should be conveyed by the wires in the kaanta metal tubes to the storage vessel. The crystal will absorb and fill the vessel with the power. In front of the storage vessel a five feet long, 3 feet high circular vessel should be installed. It should be covered all round with the bark of vaari-vriksha.

Always water will be flowing in it. So instead of water, water skin is indicated. It will give the vessel the effect of water-immersion. Then in that vessel glass cups containing the decoction of shikhaavalee or lead-wort? or achyranthes aspera?, 18 parts of ayaskaanta or loadstone? or steel acid?, and 12 parts of vajrachumbaka acid, should be placed. Then power should be drawn from the storage vessel through the wires inside the glass-covered tube, and 4 wires with glass wheel key be let into the acid vessels. Then from the bottom of the vessels 2 wires fitted with keys should be taken in a right circle to the front of the smoke-outlet stambha or pipe, and attached to the wires inside the bhujyu metal tube. The wires should also be connected to the keys of the electric friction crystals in the dhoomodgama stambha or pillar, as also to the key in the stambha. Thereby electricity will be spread in all parts of the vimaana. Therefore the vidyud-yantra or electrical machine should be installed in the left side of the vimaana.

Vaata-prasaarana Yantra
Air Spreading Machine.

Kriyaasaara says:

In order to enable the vimaana to ascend, vaataprasaarana yantra is necessary.

Therefore it is now being described. It should be made out of vaatamitra metal only.

Lohatantra describes vaatamitra loha. 13 parts of rasaan-janika or extract of Indian berbery, 27 parts of prabhanjana, and 37 parts of paraankusha, should be filled in sarpaasya or serpent-faced crucible, placed in chakramukha furnace, and with the aid of vaaranaasya bhastrika or bellows, heated to 216 degrees. Then filled in the sameekarana yantra or churner, and next poured out and cooled, it will yield vaatamitra loha, or air-companion metal.

First the foot plate, then the naala-stambha or tubular pole, air pumping wheel with keys, air attracting bellows-like mechanism, and mechanism for contracting and expanding the mouth, out-flow and inflow tubes with keys, covering for the yantras, wind pipes, vaatodgama pipe, bhastrikonmukha, vaatapoora-keelakas, vaata nirasana pankha keelaakas, or air-expelling fan keys, these 12 are the organs of the yantra.

The Peetha or foot plate.

The peetha should be 6 feet long, 1 foot thick, square or round, with two spots on the northern and southern side of it for erecting three- wheeled tubular poles.

The 3 wheeled naala stambha is described in "Yaana bindu":

Three feet long and 8 feet high tubular poles should be fixed on 2 sides of the peetha or foot-plate. At the foot and the middle and the top of the pole three openings should be provided for fixing 3 wheels.

In the pole should be fixed tubes, one foot wide and 2 feet long, for drawing in air, and wheels 1 foot wide with teeth as in hack-saw, revolving both ways, be fitted to the tubes. The vaata-pooraka or air-filling naala should be fixed in the middle of the wheel. By turning the fly wheel, the wheel will turn, making the naala move up and down sucking in air. The air pumping wheel keys should be thus fixed in the two poles. The keys at the mouth of the bellows should be connected to these keys.

Bhastrikaa-Mukha-Yantra
Bellows' mouth mechanism

Taking pig-skin, duly cleaning it with putrajeevi or wild olive oil, boil it for 3 days, wash it with clean water. Smear it with gajadantika oil frequently exposing to sun for 5 days, and fashion out of it a 6 foot bellows, three feet wide at bottom, 4 feet wide in the middle, and 1 foot wide at the mouth. Two keys working conversely to each other should be fixed at the mouth. A stick should be inserted between them. The two keys should be capable of being put into quick motion, or left at rest. By turning the keelakas the piston rod is moved, and from its speed, the bellows' mouth also will start in motion, and also the vaataakarshana naala. By putting the naala at the mouth of the bhastrika or bellows, quick air entry from inside the mouth will occur. By starting all the keelakaas in all the centres there will be airflow in the three wheel tubular stambhas. By turning the keelakas with

20 heat-degree force, in the naala stambhas air will rush out with
100 shaker speed. From the bellows' mouth also air will blow
with 2000 prenkhana or shaker speed. And these air flows will
speed the motion of the vimaana. Therefore in front of the vaatodgama yantra 12 such yantras should be installed on the four
sides, 3 on each side. And aavarana or covering should be provided for them according to their measurements. And 12 naala
stambhas, 3 feet wide and 12 feet high, should be prepared, and
fixed on the top covering of the yantras, for the air to flow out.
From each stambha air will blow with 2600 prenkhana speed.
The yantras are individually prescribed so that some may rest
when not required. The high flight will be helped by these machines. Having thus described the individual sources of air supply for the vimaana, we shall now describe the Brihat-stambha
or main mast.

It should be 4 feet wide and 30 feet high, and called vaatodgama naala stambha. It should be erected centrally amidst all
the yantras. The bhastrikonmukha yantras should be fixed at the
foot of the stambha so that the air flows from the yantras could
pass into the stambha. The wind-naalaas or pipes should be
connected to the stambha-moola fitted with keys. At the opening of the naala-stambha at the top on the 8 inch wide opening
a vessel one foot high and 3 feet wide should be fixed. The wind
from the stambha or tunnel will pass out through it in wavy bil-
lows. The dhoomodgama yantra or smoke pipe should, be provided with triple keys or fixtures, for the expulsion of smoke.
and blowing in of air. By operating those keelakas the supply
of smoke and air could be controlled according to need. Wind
expelling fan wheels should be put in, so that by their quick motion the motion of the vimaana could be facilitated.

Vimaana-aavarana-nirnaya
Covering of the Vimaana

Covering the dhoomodgama yantras and kudyaas or side
walls, as in the case of the Shakuna Vimaana, the covering of
the Sundara vimaana should be done by raajaloha only. The
covering should accommodate the number of partitions or

booths required as in Shakuna Vimaana. The location of the 32 component yantras should be determined. In the centre of the booths for locating the four-faced heat machinery, a thirty feet square area should be set apart.. There the four-faced heat yantra should be erected.

Says Yantrasarvasva:

The chaatur-mukha owshnya yantra should be made out of kundodara metal only. Kundodara metal is defined in Lohasarvasva. Soma, Kanchuka, and shundaala metals in the proportion of 30, 45, and 20 to be taken, cleaned and filled in padma crucible, placed in chhatramukha furnace, and with vaasukee bellows heated to 716 degrees, aa-netraanta, and poured into the yantra for cooling. A blue, fine, light, alloy, capable of bearing 2000 degree heat, and which cannot be blasted even by shataghnee and sahasraghnee canons, and very cold, is kundodara alloy. With this alloy the owshnyaka yantra should be fashioned.

Yantraangas or parts of the Machine.

Peetha or foot-plate, smoke container kunda or vessel, water container, fire oven, turret covering, covering of water container, twin wheels for projecting and restraining smoke, window rods, padmachakras or wheels, aavritta chakra keela, heat indicator, speedometer, time clock, ravaprasaarana keelaka naala or sound transmitting instrument, antardandaaghaata naala, air-bellows, long sundaala pipes, twin copper pipes, air dividing wheel keys, these 18 parts constitute the ooshmyaka yantra.

The peetha, tortoise-shaped, should be 25 feet long and wide. At peethaadi or starting end should be fixed the agni-kosha or fire place, the water vessel in the middle, and the smoke-container should be fixed at the other end.

The 3 koshaas are explained by Budila:

Ravi or copper, manchoulika, and tigma in equal parts should be mixed with kundodara metal, and be made into 3 inches thick

pattika or flats. One pattika should be fixed on the peetha. In
the fire place, kendra on the peetha a 4 feet long 6 feet high
fire-place should be made. For stocking coal or wooden billets,
a sort of walled table should be formed. Next a triangular fire-
place should be formed, with rods at the bottom for the ashes to
fall down. In between the 2 parts the flat sheet should be fixed,
fitted with keelakas or hinges for moving the peetha as desired.
Three keelakas should be fixed at the fire place, one to fan the
flames and straighten them, one to moderate or stimulate the
flames, and one to distribute the flames evenly. A naala or pipe
should be fixed on the fire kosa. Another pipe with wiring, is
to be fixed at the end of the fire-kosa pattika with a smoke
transmitting pipe which will convey the smoke of the fire- place
to the jalakosa or water container. From the fire kosa to the
covering of the water kosa water pipes should be adjusted. In
the water kosa enclosure the heat will rise to 5000 linkas in these
tubes. The heated water will then give out hot smoke.

The size of the jalakosa or water container is 8 feet. Three
triple-wheeled naalas or pipes should be fixed in the jalakosa:
one to restrain the heated smoke from the water, one to amass
the smoke, and one to lead the smoke into the dhoomakosa
or smoke- container. The Dhoomakosa should be 6 feet wide
and 4 feet high. In order to fill the kosha with smoke, necessary fittings should be provided. Above the jalakosa a dome-like
covering should be erected. It should be provided with fittings,
for folding up and opening out. To the front of the smoke container, two pattikaa wheels with holes should be fixed in order
to let out the smoke or to restrain it. In order to operate the
wheels two bhraamanee keelakaas or revolving switches should
be provided. To the east of the Dhooma-kunda, 8 inches long
window bars should be fixed with one inch spacings. Then in
front of the yantra, in the middle, at top, at bottom, and on both
sides, twin padmachakra keelakas should be fixed for spreading
the smoke or restraining it. For storing the wood or coal a hole
11 feet wide should be arranged. The door covering it should
be provided with needful fittings. To the north and south of
the keelaka the heat-measure and speedometer should be fixed.
Above them the timepiece. To the south, a telephonic device

called ravaprasaarana or sound ringer, which will give alarm
with 1212 sound wave speed, and which gives warnings for the
plane's moving, halting, speeding, overspeeding, and danger im-
minence. An equipment with 5 holes giving 5 different sounds
to indicate the above should be installed. On either side of the
above, two 6 inches wide, 26 feet tall, Aaghaatha-naalas or pipes
should be fixed. Between them two 5 inches thick metal rods are
to be adjusted. At the foot, middle, and top of the naalas revolving wheel keelakas should be fitted. By their revolving, the rods
will strike each other. That will increase the speed of the plane.
On the top of the naala pipes, air bellows with fittings should be
fixed. Thereby the air force in the naalas will shoot up, and the
speed of the vimaana will double. Then on the four sides of the
heated smoke kosha or container, shundaalas or elephant-trunk-
like pipes should be fixed with wheeled keys as in vaatodgama
yantra. By filling the shundaalas with the smoke and turning the
keys as required, the movement of the vimaana in one direction
or another, its gaining height and speeding out or halting, will be
facilitated. Keys should be adjusted so as to make the shundaalas coil down like a water hose or keep erect. Two pipes made of
3rd division copper should he wound round the agnikosa, water
kosa, and smoke kosa, or fire, water and smoke koshas, in order
to absorb the excessive heat in them.

In order to part the wind in front of the vimaana, vaata-
vibhajana chakra keela or wind-dividing-wheel fittings should
be fixed.

Having thus prepared the chaaturmukhoshmyaka yantra, or
four-faced heating machine, it should be installed in the centre
of the vimaana. By the air, smoke, and heat of the yantras below, the ascent and flight of the vimaana will be facilitated.

Regarding the speed of the vimaana, we have to consider
the speed of smoke and other accessories mathematically, and
conclude the possibility of the speed of the vimaana. The speed
of the smoke from dhooma yantra is 2113 linkas. The speed
of wind from the air blowing machine 2500 linkas. Wind from
the naala-stambha blows at the speed of 600 linkas. This is the

speed of the forces from the 3 machines on the peetha. Of the forces from the upper portion of the vimaana, from the chatur-mukhoshmyaka yantra, heat force of 3400 linkas emanates. By the four-faced heat yantra, and by operating the keys of the shundaalas, and the force of the wind, smoke and heat machines, the vimaana would be capable of a speed of 400 yojanas or 3600 miles.

This is Sundara Vimaana, and it has been described after consulting ancient works, and according to my humble capacity, says Maharshi Bharadwaaja.

RUKMA VIMAANA
"Atha Rukma Vimaana Nirnayaha"

Next the principles of Rukma Vimaana.

Maharshi Bharadwaaja:

☞ **"Rukmascha" Sootra 1.**
"Rukma too"

Bodhaananda Vritti:

This vimaana is of golden colour. Therefore it is called Rukma vimaana, Rukma meaning gold. The Rukma should be made out of Raajaloha only. By duly processing, Raajaloha can be made to assume golden colour. That metal should be used for the vimaana.

"Yaana-Bindu" says,

"After first producing golden colour for Raajaloha, the vimaana should be formed."

"Varna-sarvasva" mentions the colouring process:

Praana-kshaara or ammonium chloride 4 parts, wild Bengal gram 32 parts, shashakanda (or lodhra?) benzoin? 18 parts, naaga or lead 20 parts, sea-foam 16 parts, maakshika or iron pyrites 6 parts, panchaanana or iron 20 parts, paara or mercury 15 parts, kshaara-traya or 3 kinds of salt: natron, salt-petre, borax, 28 parts, panchaanana or mica 20 parts, hamsa or silver 17 parts, garada or aconite 8 parts, and panchaamrita or 5 sweets—curds, milk, ghee, sugar, honey, these should be filled in the melter, and after boiling, and drawing the liquid through two outlets, fill in the crucible and place in furnace, and blow to 800 degrees' heat, and then transfer it to the cooler.

That will be Raajaloha, pure, golden-coloured, tensile, and mild. The vimaana, made out of this loha or alloy, will be very beautiful and delightful.

The Peetha

The peetha or ground plate of the Rukma vimaana should be tortoise-shaped, 1000 feet long, and 1 foot thick, or any other desired size. On its eight sides, 20 feet long spaces should be fixed underneath the peetha. At each centre fixtures like birds' beaks should be attached with revolving keelakas. Then double iron-balls or wheels, in couples, should be fixed in each of the 8 centres.

Ayas-chakra

Lalla gives the form of ayaschakra-pinda:

12 feet long and wide, and 8 kankushtas in weight, they should be made round like a grind-stone. They should be inserted in the beaks at the 8 centres. From each chakra-pinda up to the electrical generator chain wires should be connected with switches.

Batinikaa-Stambha
Or Button-switch pole

One foot wide and 4 feet high poles should be fixed. They should have switches wired up to the electric pole. 8 inches wide wheels should be fixed in the middle of the pole, on either side, with wires. From the electric pole chain wires should enclose the wheels and be fixed in another pole with inside hinges. On the top of the poles should be fixed goblet shaped cups with button-switches like half- blooms with wheels and keys, so that on pressing the button with the thumb the wheels in the other pole will revolve from electric contact. Then the wheels in the electric pole will also revolve, producing 5000 linkas of speed.

Flying

Due to this electrical force, the ayah-pinda wheels beneath the peetha will beat against it and make it rise and move upwards. And by moving the switches of the wheeled poles above the peetha, the poles will revolve with speed, and accelerate the speed of the vimaana. By the concussion of the wheels underneath, and the action of the poles above, the vimaana will move upwards and gain height and fly with dignity.

Electric tube wheels aiding flight:

Above the peetha, naalas or tubes should be fixed at 1 foot intervals. On both sides of each naala toothed wheels 2 feet wide and 1 foot high should be fixed with proper keelakas. Taking electric wires through the keelakas, and passing over the wheels and reaching the foot of each naala, they should be attached to wheels 3 feet wide and 3 feet high. In the midst of 20 naalas a pole should be fixed in the centre.

Narayana says:

Preparing a pillar 4 feet wide and 4 feet high, and making a 2 feet opening in its middle, fix keelakas at the top, middle, and lower end of the opening. Two keelakas with 6 wheels, with glass coverings, with wires, and naala and leather covering should be fixed at the lower end for attracting electricity. In the middle part of the opening, for transmitting the current, a five-faced keelaka should be fixed, with 5 wheels, glass covering, 2 naala tubes, two wires, attached to 3 rods, and vessel containing veginee oil. By the flow of the current the wheels in the upper end should be made to whirl by properly adjusting keys. In front of the opening a big wheel should be fixed with gumbha keelakas. Similarly wheels should be fixed at the foot of each pillar. On top of them a four inch wide pattika or flat band should be adjusted commencing from the samsarga key chakra up to the front of the electric yantra. By operating that key, power will flow through the wires, and entering the key at the foot of the pillar set the wheels in motion. On the motion of the big wheel the sandhi- wheels in the naala-dandas will also revolve with speed, and the current will enter the 5 faced keelaka, and entering the oil vessel it will gather force, and passing through the 2 naalas, set all the wheels in the pillar in forceful motion, generating 25000 linkas speed, which will give the vimaana 105 krosa or nearly 250 miles speed per ghatika, or 24 minutes.

Having dealt with the mechanism for setting the vimaana in motion, we now consider the mechanism for giving direction to the vimaana in its course. In the 8 diks or directions of the peetha, pillars made of mica and shining like panchakantha, 2 feet thick and 15 feet high should be fixed at intervals of 10 feet. On the pillars should be built the passenger seating arrangements, and booths or locations for the machinery, as in the case of the Sundara Vimaana. The pillars should be made of mica only.

Its production is given in Kriyaasaara:

Shaara-graava or lime 25 parts, kshwinkaasatva or iron-sulphate 30 parts, gunja or wild-liquorice 28 parts, tankana or borax 12 parts, roudree moola 8 parts, chaandree or kantakaari.... solanum xanthocarpum flower salt 2 part, purified shoonya or mica 100 parts,, to be filled in koorma crucible, and heated in paadma furnace with blower to 800 degrees, and then poured into the cooler, will yield mica alloy most useful and attractive. Fashioning the pillars or walls or partitions and booths, and fixing the mechanisms for turning, circling, diving, and manoeuvring, in the fore and middle and aft of the vimaana, it could be moved in any direction as desired.

Lallaacharya says:

In order to make the vimaana change its course from one path to another or one direction to another, revolving keelakas should be fixed on the eight sides of the vimaana. Two keelakas should be made, purva and apara, or right side and left side. They should be fitted together.

By operating it, the vimaana could be made to change its course one way or another. In order to operate the keelaka, at the peetha moola, on the 4 sides crescent shaped naalaas or tubes, 2 feet wide and 2 feet high should be fixed. 4 inches long metal rods should be fixed inside the naalaas on either side. One foot wide and 1 foot high wheels should be fixed in them. They should be wired all around. Such crescent naalas should be fixed on the 4 sides of the peetha. In order to set the wheels in the naalas in motion big wheels should be fixed at the beginning, middle, and end of the naalas. By turning the top wheel with speed the wheels inside the naalas will revolve. That will force the keela-shankus to twist round so as to force the vimaana to change its course in the required direction.

TRIPURA VIMAANA

Maharshi Bharadwaaja:

☞ **"Tripurothha." Sootra 2.**
"Next Tripura."

Bodhaananda Vritti:

Having explained the vimaanas commencing from Shakuna to Simhikaa, Tripura vimaana will now be dealt with.

This vimaana has 3 enclosures, or aavaranas or tiers. Each aavarana is called "Pura." As it consists of 3 aavaranas it is called "Tripura" vimaana. It is operated by the motive power generated by solar rays.

Narayana also says:

The vimaana which naturally can travel on land, sea, and in the sky by alteration of its structure is called Tripura Vimaana.

It has got 3 parts. The first part can travel on land. The second part can travel under and over water. The 3rd part travels in the sky. By uniting the 3 parts by means of keelakas, the plane can be made to travel in the sky. The plane is divisible into 3 parts so that it might travel on land, sea, or air. The construction of the 1st part is now explained. Tripura vimaana should be made out of Trinetra metal only.

Trinetra loha is explained by Shaakataayana:

Jyotishmatee loha 10 parts, kaanta-mitra 8 parts, vajramukha loha 16 parts, these 3 to be filled in crucible, then adding tankana or borax 5 parts, trynika 7 parts, shrapanikaa 11 parts, maandalika 5 parts, ruchaka or natron 3 parts, mercury 3 parts, then filled in crucible in padmamukha furnace and heated to 631 degrees with trimukhee bellows, the resulting liquid, if poured into cooler, will yield a metal, shining like peacock feather, unburnable, unbreakable, weightless, impregnable by water, fire, air and heat, and indestructible.

With that metal the peetha should be prepared, of any desired size. The following is given as an example. It may be 100 feet wide and 3 feet thick, round or square. Leaving 20 feet on the western side, at intervals of 10 feet 80 spots should be marked for wheeled boats. 80 feet long, 3 feet wide, 5 feet high boat shaped dronies or containers should be fixed on the marked lines. Three feet wide openings should be made in the top of the dronies, so as to raise the wheel inside them quickly and cover them underneath. There should be fittings which enable the wheels to be lowered on land, and raised and covered underneath when going in water. The wheels should have axle rods with fittings to attract electric power. The axle rods should be 2½ feet long and 1 foot thick. The wheels should be 3 feet wide and 1 foot thick, have, 5, 6, or 7 spokes, fixed in the rims, and covered with musheeka up to 4 inches from the edge. Holes with glass coverings should be made in all the wheels. These 12 wheels, or 8, or 6, or 4, should be fixed inside the boat-like structure. For transmitting power wires made of so-makaanta loha should be fixed in the holes made in the wheels. In the middle of each wheel electric aaghaata keelakaas should be fixed, and in them chhidraprasaarana keelakas. Over all the chakradronee boats, copper wire pairs should be fixed on both sides, and in the joints of the wheels. Rods should be attached to the wires so that power could be drawn from the wires and passed to the top of the wheels. And power should be passed to the wires underneath the wheels. In climbing hills, and going down slopes, by adjusting the power at the top or the bottom of the wheels, smooth progress is made possible. By adjusting the necessary keelakas it is possible to accelerate the speed, or in going down, to restrain the flow of the current, and put brake on excess speed.

For attracting power from the generator a naala or pipe with wires should be fixed at the front of the peetha through 5 faced wheel keelakas, and the wires should be connected to the fittings at the top and bottom of the wheels, with glass cups.

In order to put covering over the boat formations, pillars should be fixed between each boat line, and covered with mica sheets, as per architectural rules.

Maharshi Bharadwaaja:

☞ **"Shuddhhaambaraattadhhi." Sootra 3.**
"Out of pure mica alone"

Bodhaananda Vritti:

The vimaana should be made out of pure mica alone.

Mica is described in "Dhatu sarvasva ". There are four kinds of mica, white mica, red mica, yellow mica, and black mica. The white mica has 16 varieties. Red mica has 12 varieties. The yellow mica has 7 varieties. And the black mica has 15 varieties. Thus there are 50 varieties in all.

Shownakeeya also says:

We shall now describe the nature of abhraka or mica. They are of 4 castes, like brahmin, kshatriya, vysya, and sudra. They are of 50 varieties. The brahmin mica has 16 varieties. The kshatriya mica has 12 varieties. The vysya mica has 7 varieties. And the sudra mica has 15 varieties, totalling 50 in all. Their names are as follows. The brahmin mica varieties are ravi, ambara, bhraajaka, rochishmaka, pundareeka, virinchika, vajragarbha, koshambara, sowvarchala, somaka, amritanetra, shytyamukba, kuranda, rudraasya, panchodara and rukmagarbha. The kshatriya varieties are shundeeraka, shambara, rekhaasya, owdumbara, bhadraka, panchaasya, amshumukha, raktanetra, manigarbha, rohinika, somaamshaka, and kourmika. The vysya varieties are krishnamukha, shyaamarekha, garalakosha, panchadhaara, ambareeshaka, manigarbha, and krownchaasya. The shoodra varieties are gomukha, kanduraka, showndika, mugdhaasya, vishagarbha, mandooka, thailagarbha, rekhaasya, parvanika, raakaamsuka, praanada, drownika, raktabandhaka, rasagraahaka, vranahaarika.

Out of these, pundareeka from the 1st class, rohinika from the second, panchadhaara from the third, and drownika from the 4th class are good for use in constructing the vimaana. These should first be purified as per rules.

The process of purification is given in "Samskaara Ratnaakara": skandhaaraka or salt of roitleria tinctoria?, shaaranika or rubus salt?, pinjulee or yellow orpiment?, cowries, borax, kaakajanghaa or wild liquorice?, moss, rowdrikaa, salt-petre, douvaarika, shambara or benzoin, and phosphorus. These should be separately filled in the smelter. The decoctions should be filled in glass vessels. The mica is to be purified with each one of these.

The mica is to be powdered, put in skandhaavaara acid in smelting vessel. It should be boiled for 3 days in fire, and for 3 days in electric heat. Then take the liquid and put it in a bronze vessel, pour in shaaranika acid and keep it in sun for 3 days. Then add pinjulee acid and keep buried in earth for 5 days. Afterwards add cowri acid, and boil in bhoodhara yantra for one day. Then add mustard, and adding borax acid and burning arjuna, myrabolan wood, place it in brown-barked acacia cinders for 3 days. Then add wild liquorice acid and expose it to the full moon rays on the 14th and 15 days. The mica is to be then taken out and washed in hot water. Then add wild corn, and pouring in moss acid place it under earth for 6 days. Then take out the mica, add roudri acid, place the vessel in a big fire-place, and burn in 64 feet of dried cowdung. Next taking out the mica put it in sesamum oil for 1½ days, and expose to the sun from morning to sundown. Then take out the mica, wash it clean, put in bronze vessel with salt-petre solution with dattoori or yellow thistle seeds, place it in a heap of burning kundalee or mollugo stricta leaves. Then take out the mica, add dourvaarika acid and bake for a day with hay-fire. Then put the mica in benzoin acid for 3 days. Next add one-fourth as much of camphor, and placing it in the churning machine, churn for a day. Then placing it in Simhaasya crucible cook with boiling water. Add ranjaka or phosphorus acid, 3 palas or 12 tolas of tankana or borax, 12 tolas of lime, 4 tolas of soorana root or tacca, karkotaka 20 tolas, vrishala or onion 28 tolas, koorma-tankanaka 8 palas or 32 tolas, rouhinaka or red sandal 40 tolas, shambara 80 tolas, muchukunda 12 tolas. These cleaned and filled in the crucible, and placed in simhamukha furnace filled with charcoal, and melted with 800 degrees heat will yield a metal shining like a precious stone, very light, unbreakable, unburnable and indestructible.

With that the vimaana is to be constructed.

We shall now consider the parts of the vimaana: 2 feet thick and 3 feet high pillars, painted in different colours and adorned with pictures, should be prepared, and 80 of them should be fixed in the spaces between the boats. On the pillars 10 feet wide pattikas or sheets, and of the same length as the boats, should be fitted with screws, and two-faced hinges.

In order to accommodate crew and passengers of the vimaana, and store luggage, rooms and partitions should be constructed with decorations. In order to provide secrecy, doors should
be provided as also ventilators. Revolving wheels with necessary
fittings and switches should be fixed so that by putting on a
switch the rooms would revolve. Wheels should be fixed in the
lanes between the boats. Air-pipes with wheels should be fixed.
In order to ensure supply of air, tubes with wheels, and bellows
with wide mouths, leaving 20 junctional centres, should be fixed.
In the front, two faced tubular wheels should be fixed to dispel
the air downwards or upwards or side ways, at 30 feet intervals
from the aavrutta or enclosed pradesha of the vimaana. At the
bottom of the vimaana metal balls with chain-wirings should be
fixed for operations in the course of flight.

The 1st floor will be 7 feet high, with the roofing duly fixed
with nalikaa-keelakas with 10 feet intervals. With 20 feet interval
in the middle, wires with beaked ends should be attached to
each keela. The fittings should be such as to enable opening
and shutting like an umbrella. The cloth covering like a tent top
should cover the entire floor.

The second aavarana should be made of trinetra metal.

Maharshi Bharadwaaja:

☞ **"Taduparichaanyaha." Sootra 4.**
  "Another above it."

Bodhaananda Vritti:

Having described the first floor above, now the second floor
is being described. The second floor should be slightly smaller
than the first floor. If the first floor is 100 feet wide, the second
should be 80 feet wide. The floor should be 80 feet wide, and
3 feet thick, and made of trinetra metal. Its fittings should be
like those on the first floor, and be duly connected with electric
wiring from the generator.

In order to take the vimaana through water, first the wheels at the bottom used for land route should be drawn up, and in order to prevent water coming up, the bottom should be completely covered up with ksheeree-pata or milk cloth. Four inches thick metal rods, 12 inches long, to which wheels 1 foot wide and ½ foot thick, and shaped like frog claws, are fixed, should be adjusted on both sides of the dronee or boat lines. Similarly in the front portion of the vimaana, on both sides two such wheeled rods should be fixed in order to divert water, By switching on power the main wheels will revolve, making all the wheels revolve, and expelling water, and aiding the progress of the vimaana forward.

For the supply of air inside, on the sides of the 2nd floor, should be fixed, air pipes 6 inches wide and made of ksheeree pata or milk cloth, cleaned with acid, from the partitions in the 1st floor upto the top of the vimaana, their tops being covered with revolving metal covers, with air sucking pumps worked by power. The air so pumped into the pipes will fill both the second and 1st floors, and provide air comfort for the crew and passengers of the vimaana.

Above the roofing of the two floors all round, spreading out and closing up keelakas should be fixed. So as to separate the floors, foldable chain fittings should be fixed at 10 feet intervals. Wires from the electrical generator should be connected to the fittings, so that by their operation the floors will be separated, and the separated floors simultaneously move on land and in the air.

In the 2nd floor also cabins, partitions and seating and doors and windows should be constructed as attractively as in the first floor. The enclosing walls of the floor should be 7 feet high from its peetha, and half a foot thick. In order to draw electric current from the third floor two poles should be erected in the back room with transmitter from which wires will pass the current to the various fixtures on the floor.

At the front of the vimaana a mast should be erected. At its

foot two bells made of bronze should be fixed in order to indicate time to the crew and passengers. In every room on the floor
alarm chains, as in railways, should be fixed so that the occupants
may call for help in times of danger. On hearing the call the crew
will rush to the room and attend to the requirements of the passengers. Sound transmitter, image transmitter, direction indicator, time-piece, and cold and heat gauges should be installed on
either side of the floor, with necessary cable connections.

Then in order to protect against excessive wind currents,
storms, and heat-waves, three machines should be installed at
the back, on either side, and on both sides of the turret.

They are described in "Yantra Sarvasva" as three-faced air
protection yantra, solar-blaze conditioning yantra, and rain
storm protection yantra. Their construction is given here as per
shaastras.

First, three-faced air force reducing yantra.

It must be made of Vaaruna Metal:

Vaaripanka, vishaari, borax, jaalikaa, mango, vishodara, vaar-
ipanchaka, kshaarasaptaka, kshona, manjula or madder root,
godhara, vaarunaasyaka, paarvana or chlorodendrum phlomoides, aruna, kaakatunda, bhoodhara, vaarunaabhraka, natron,
kundaaleemukha, lodhra or benzoin, varikudmala or water
flower, shaarikaarasa, panchabaanasahodara, lead 5 parts, soorana or tacca, honey 8 parts, vaata, kankanikodara, Sunda, anjana
or eye-black, kukkutaandaka, khaadira or brown-barked acacia,
loddhruka, simhikaa-mukha, koormajangha, and masoorika or
lentil, all these to be cleaned, and filled in crucible, placed in
padmamukha furnace, and heated to 700 degrees with 5 faced
bellows, poured into equifying yantra and churned, will yield a
light, smoke-coloured, impregnable vaaruna metal.

Then it is to be purified, according to "Kriyaasaara." First,
place it in shundeera acid (great-leaved laburnum?) and boil for
3 days, and then with kuttinee yantra beat it into flat pattis, make

thick decoction of soorana root or tacca, and smear it to 1 inch thickness on it and heat it for 3 yaamaas or 9 hours. Then mritsaara, vaagura, opium, should be boiled together for a day. The concoctions will become red like lac. The metal patti should be smeared with it and heated in the taapana yantra for a yaama or 3 hours. Then keep it in the sun for a day. Then kantaka or small caltrap, heranda, dhavalodara, and chaaraka, and gingelly should be mixed together, and the oil extracted. The metal should be smeared with it and kept in the sun for 3 days, and then heated in the sun for a day. Then paste the gum of kankola or cubeb pepper 1 inch thick, and stick into it thumb-sized vaatakuthaaraka manis, place in furnace of brown- barked acacia and cool for 9 hours. The metal will become like diamond.

Out of this a cover should be made for the vimaana, with necessary fittings for spreading over and folding up, connected with electric wires drawn from inside the vimaana. The charge of electricity will permeate all over, as well as the manis on the pattika. Three serpent-faced keelakas should be fixed. These will suck in the fierce wind as it blows, and belch it out to the upper regions, so that the wind force on the vimaana will be curbed, and danger therefrom averted.

The rain storm protection yantra should be made of crowncha metal. Says "Kriyaasaara", The metal that can destroy the dravapraanana force of water is krowncha loha. Therefore the varshopasamhaara yantra should be made out of that alone.

Krowncha loha is described in "Mantra Sarvasva" as follows: Jyotirmukha or rose-coloured red-wort 8 parts, tryambaka or copper 11 parts, humsa-tunda 12 parts, camphor 7 parts, tankana or borax 8 parts, sand 4 parts, choorna or lime 12 parts, owrwaara or cucumber?, ruruka 5 parts, patola or snake-gourd 27 parts, and vaardhyushika or sea-foam 1 part, these to be cleaned and placed in crucible, and heated in padma furnace to 512 degrees with 3 faced bellows, poured into churning yantra, and then cooled, will yield, a metal, honey-coloured, light, strong, rain-storm antidote, and heat-impregnated. Extracting oil from the seeds of basil, rukma or yellow thistle, punkha, red wort, trijataa or bael, and pancha-kantaki or 5 thorny trees, the metal should be smeared and heated. The metal is to be made into pattis with kuttinee yantra, make pipes out of them 3 feet wide of the same height as the vimaana, and fix them properly all around. In front of the vimaanaa-avarana also 3 feet high pipes should be fixed with keelakas or hinges. The pipes should be smeared with chana or gram decoction 1 inch thick. On that vajragarbha decoction or triangular spurge milk should be smeared thrice, which will make it hard as diamond. On the pipes, at 12 inches intervals, sinjeeta vajra should be smeared and heated by fire. Then thumb-size panchaasya manis which will counteract the effects of water, should be imbedded on the smeared pipes. Then the pipes with proper fittings at both ends should be fixed on the 8 sides of the vimaana. Wires proceeding from the electric generator should be taken through glass tube and connected to the pipes. When the current passes through them to the panchaasya mani, the concentrated force in it blending with the electric force will fiercely oppose the forces of the rain storm and disturb the atmosphere so as to dilute and weaken the storm, and render it ineffective. Therefore the varshopahaaraka yantra should be fixed on the vimaana.

Sooryaathapopasamhaara yantra or the burning-sun protection machine:

It is to be made out of the aathapaashana loha. It is explained in Kriyaasaara: Aatapaashana loha protects against burning sun. Therefore Aatapa samhaara yantra should be made with that metal. "Lohatantra" describes that metal. Owrvaarika, kowshika, gaaruda, soubhadraka, chaandrika, sarpanetra, sringaataka, sowmyaka, chitraloha, vishvodara, panchamukha, virinchi, these twelve metals should be put in equal parts in padma-moosha crucible. Borax 7 parts, chowlika 5 parts, cowree salt 6 parts, kunjara 12 parts, sand 9 parts, camphor 4 parts, cardamom 16 parts, powshnika 10 parts, should be added to them, and placing it in nalikaa furnace heated to 725 degrees with mooshakaasya bhastrika bellows. Then the liquid should be put in, the mixing machine, and afterwards poured into the cooler. The resulting alloy will be light, orange coloured, heat proof, and unbreakable, for the making of sooryaathapopasamhaara yantra, after being duly purified, says Yantrasarvasva.

Kriyaasaara explains its purification:

Ashwaththa or sacred fig tree, mango, plantain, aala or banyan, baadava or peepul, trimukhee, trijata or bael, gunja or wild liquorice, sherinee, onurberah, patolika or snake gourd, the bark of these trees should be powdered, should be filled in vessel with 10 times as much water, and boiled down to one-tenth measure.

Then taking the 11 kinds of salts, bidaa-lavana or table-salt, syndhava or rock-salt, oushara or saline earth, budila salt, maacheepatra salt or solanum indicum?, praanakshaara panchaka, or 5 urine salts or ammonium chloride? and saamudra or sea-salt, these eleven salts, should be placed in dravaakarshana yantra or dehydration machine and boiled. Taking the previous decoction, add half as much this decoction, put the aatapaashana metal in it and boil for 5 days, then wash with water, and anoint with honey, and place in hot sun for 3 days, then wash it, and use it for producing the yantra.

First pattikas should be made from the metal with kuttinee yantra, 2 feet square, or circle, and 3 feet thick. On that 3 pipes, 1 foot wide and 5 feet high, should be fixed. Three triangular glass bowls should be placed underneath the pipes. In each of them one prastha or seer of somadraavaka or white acacia juice should be filled. In each vessel a heat proof crystal of the 121st class should be cleaned with acid and placed. Then an umbrella shape 10 feet wide should be made out of the metal, and fixed so as to cover the 3 pipes, with revolving keelakas fixed half-a-foot underneath the umbrella cover. Above that 3 kalasas, 3 feet wide and shaped like cooking vessel, should be fixed. At their centre circular chaalapattikas should be fixed. Upon that three cold-diffusing crystals of the 185th number, should be fixed. On them three black mica wheels should be fixed. They should be covered with chandrikaa toolikaa or white silk cotton. On that should be placed a vessel with acid of manjoosha or madder root, in which a heat-resisting crystal is immersed. In the front part the toothed mica wheels fitted with bhraamanee-danda keelakas should be fixed. And in order to revolve that keelaka 3 wheeled keelaka should be fixed. By its motion the umbrella will revolve disturbing the heat wave. Then the heat-absorbing mica wheels will absorb the heat, which, pas-sing down to the madder-root acid, will become cold and get extinguished. And the crew and passengers will be saved from its evil effects.

The Third Floor:

In erecting the 3rd floor of the vimaana, the same procedure as was followed in erecting the second floor should be followed. Like the fixtures in the flooring of the 2nd aavarana and roofing of the 1st aavarana, fixtures should be put in connecting the roofing of the 2nd aavarana and the peetha of the 3rd aavarana. The peetha of the 3rd floor should be 5 feet less than the peetha of the 2nd floor, and be square or circular like it. The cabins, doors, walls, and furniture on the 3rd floor should be on the same lines as in the 2nd floor. In the north eastern part of the 3rd floor, a cabin should be prepared for housing the electric generator. It should be made out of somaanka loha.

Somaanka loha is explained in "Lohatantra" as follows: Lead, panchaasya, and copper, 7 parts each, Chumbaka or load-stone 9 parts, nalikaa or Indian spikenard bark, sharaanika or rubus salt?, and borax, in equal parts, to be filled in sarpamukha crucible, and placed in naagakunda furnace, filled with coal, and heated to 353 degrees with shashamukha bellows. After melting the liquid should be filled in the mixer, and after churning be poured out to cool. The resulting metal will be a fine, light, electricity-impregnated somaanka loha. Out of that metal pattikas should be made with kuttinee yantra, or hammering yantra.

A cradle-like vessel, 3 feet wide and 8 feet high, should be made out of it, and be covered with a pattika with hinges. On the eastern and northern part of the cover two holes 1½ feet wide should be made. The cradle should be fixed in the electric cabin. Below the holes, two peethas should be fixed in the cradle. Two vessels 2 feet wide and 4 feet high should be prepared. Eight goblets 6 inches wide and 1 foot high should be made, and 4 each should be placed in the two vessels, in their four corners. In the middle of the 4 goblets, a big goblet should be placed so as to contact all the four. 2 vessels covered with patties having 5 holes should be placed inside the 2 holes in the cradle cover. Teethed churners 5 inches in size, 8 inches in height, like those of sugarcane machines, 8 in number, should be placed in the 8 goblets in the two vessels in the cradle. 2 churners, bigger than these should be placed in the two central goblets beneath the two holes. Fixtures should be fixed on the central churner so that by their turning all the other churners will turn.

The procedure for extracting electricity out of solar rays is as follows. 8 naalas or tubes should be prepared out of the 192nd kind of amshupa glass. The naalas should be fixed on the 4 corners of each vessel. Panchamukhi karnikaas should be placed on them, filled with rukmapunkhaa shana, and with electric crystals in them. Covering them with the amshupaa glass cover, 5 spires should be formed on it. The top of each spire should be like an open beak, and in it should be inserted sinjeeraka crystal and amshupaa crystal. On the central spire amshu-mitra mani should be fixed. Above the 4 crystals should be fixed 4 glass tubes made of kiranaakarshana glass, 6 inches wide and 3 feet high. On them should be carefully fixed 4 feet-wide- mouthed vessels, acid cleaned. They should be filled with Rudrajataa-vaala or aristolochia indica linn. Revolving ghutikas should be placed in their centre. The ghutikaas will attract the solar rays and send them through the tubes. The crystals in the spire beaks will

suck them in. So does the shinjeera crystal inside, as also the amshu-mitra crystal. The power will be absorbed by the glass-covering, and sent to the electric crystal. Then the karnikas inside will receive it and send down to the central tube with force. When the central churner revolves the other churners also revolve. The power will enter the acid, and the crystals in it will whirl with great speed, intensifying the power force to the extent of 1080 linkas. That force should be collected by the ganapa-yantra in front of the cradle, and stored in the central storage.

The Ganapa-yantra is a machine shaped like Vighneshwara, 1 foot broad, and 3 feet high. From its head a tubular projection like

Vighneshwara

elephant's trunk, covered with glass and with wires inside should be fixed at the front of the cradle, and connected to the Ganapa image from the neck to the navel. Three-inch toothed wheels should be so fixed that a big wheel at the neck of the image, by force of the current coming through the trunk or proboscis will whirl, setting the other wheels in motion. A coil of wire should be placed in the centre. On it a sapta-shashthi shankha or conch called simhikaa should be placed, with covering made of kravyaada metal. 5 spoonfuls of jeevaavaka acid (ditamine?) should be filled in the conch, and 217 bhaamukha graamukha manis or beads should be placed inside. 5 umbrellas, 2 inches wide, should be made, and 5 sun-crystals of the size of big liquorice, should be stuck on them. The umbrellas should be fixed on the conch, with amshupa glass covering. This should attract the force of the sun rays, and pass to the crystals on the umbrellas, making the crystals and the umbrellas whirl with

fierce force of 1000 linkas, and the force passing to the acid in the conch and the crystal inside, will thence pass westwards, and could be transmitted through wires for any desired use. To measure its exact force a meter should be fixed in, along with thermometer and other needful equipments.

THE GROUND WHEELS

When the vimaana has to move on the ground, the electric current is switched on the electric motor in the hub of each wheel, thus causing the rim to revolve and move the vimaana.

But when entering water the wheels are drawn in by the movements of toothed segment and the pinion, the latter being revolved by an electric motor attached to the shaft. The openings in the bottom of the vimaana are closed by the sliding covers moved by the rack and pinion arrangement, the pinion being worked by an electric motor.

The movements of the hinged joints of the folding links will raise or lower the second floor over the first floor.

ELECTRIC GENERATOR

Two jars are placed on the peetha or stand. Each jar contains five cups filled with acids. Each cup has a churning rod with gear- wheels connected together. The wheels are revolved by hand while starting, and by the generated electric power afterwards. A darpana or mirror and gharshana manis are fixed above the gear wheels. The darpana and the manis absorb the sun's energy and transmit it to the acid cups. The acids, being churned, convert the absorbed energy into electric current, which will pass through the pancha-mukhee naala, or five-way-switch, to different points, and work the machines there.

THE ELECTRIC MOTOR

The electric motor consists of a loop of fine wire coil, with a fine wire cage in the centre. The current from the generator is brought to the wire coil through a glass tube. Suitable wheels are attached to the wire cage to connect to the churning gears of the generator or the shaft of the pinion.

The Simhika shankha on the top of the motor contains an acid and the bhaamukha-graahinee mani or crystal. Five rods with amshupaamitra manis are fitted to the top of the shankha, and toothed wheels are fitted to these rods to revolve together and rub against the inner surface of amshupaa mirror at the top. The solar power absorbed by the mirror is stored in the shankha, and given out by the bhaamukha graahinee mani to the various motors in the vimaana.

Thus concludes the description of Tripura Vimaana.

And that brings us to the end of the

WONDER MANUSCRIPT

left behind for the edification of Mankind
by the venerable mystic

ANEKAL SUBRAAYA SASTRI

whose occult powers visualised this much from the

"VYMAANIKA SHAASTRA"

section of the giant

"ENCYCLOPAEDIA OF MACHINES"
or
"YANTRA SARVASVA"

of divine sage

MAHARSHI BHARADWAAJA

RUKMA VIMANA

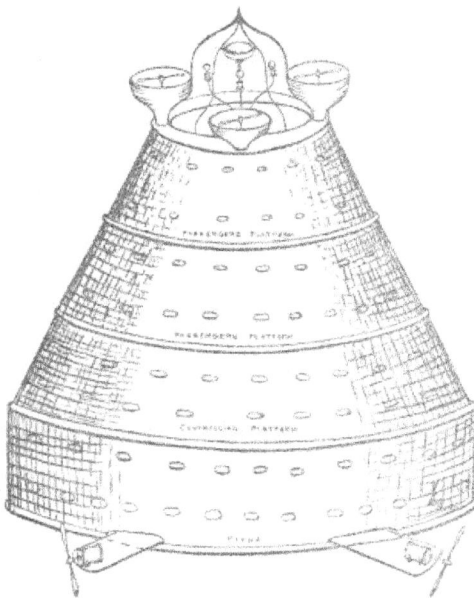

PROFILE

Drawn by
T. K. ELLAPPA,
Bangalore.
2-12-1923.

Prepared under instruction of
Pandit SUBBARAYA SASTRY,
of Anekal, Bangalore.

RUKMA VIMANA

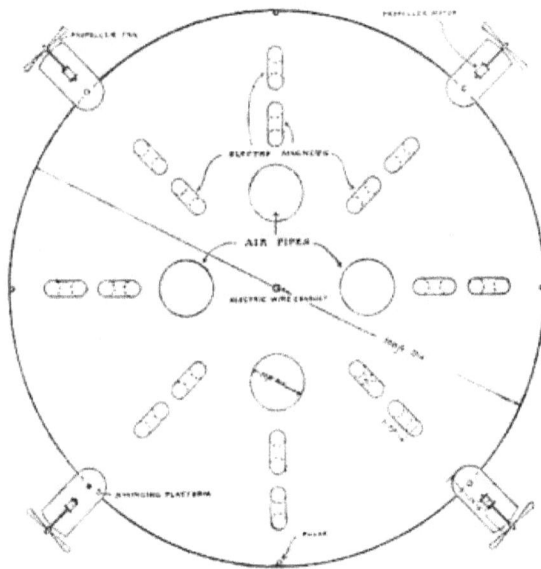

PLAN OF BASE OR PITHA

Drawn by
T. K. ELLAPPA,
Bangalore.
2-12-1923.

Prepared under instruction of
Pandit SUBBARAYA SASTRY,
of Anekal, Bangalore.

RUKMA VIMANA

Plan of Top Floor

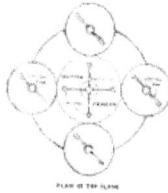

PLAN OF TOP FLOOR

PLAN OF STEERING FLOOR

PLAN OF STEERING FLOOR

Drawn by
T. K. ELLAPPA.
Bangalore.
2-12-1923.

Prepared under instruction of
Pandit SUBBARAYA SASTRY,
of Anekal, Bangalore.

RUKMA VIMANA

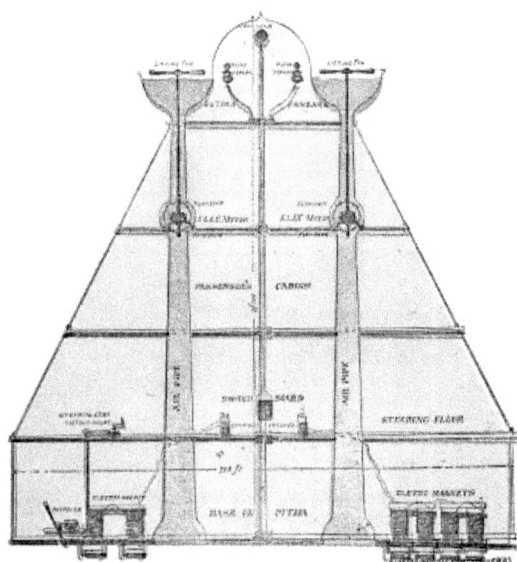

VERTICAL SECTION

Drawn by
T. K. ELLAPPA,
Bangalore.
2-12-1923.

Prepared under instruction of
Pandit SUBBARAYA SASTRY,
of Anekal, Bangalore

SUNDARA VIMANA

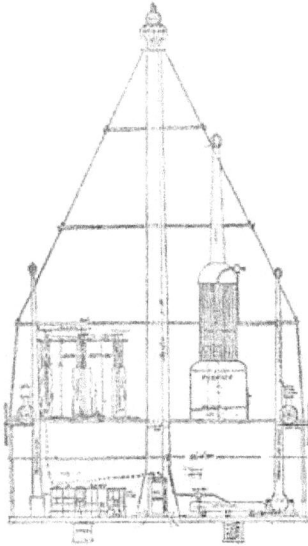

VERTICAL SECTION

Drawn by
Y. K. ELLAPPA,
Bangalore.
2-12-1923.

Prepared under instruction of
Pandit SUBBARAYA SASTRY,
of Anekal, Bangalore

SUNDARA VIMANA

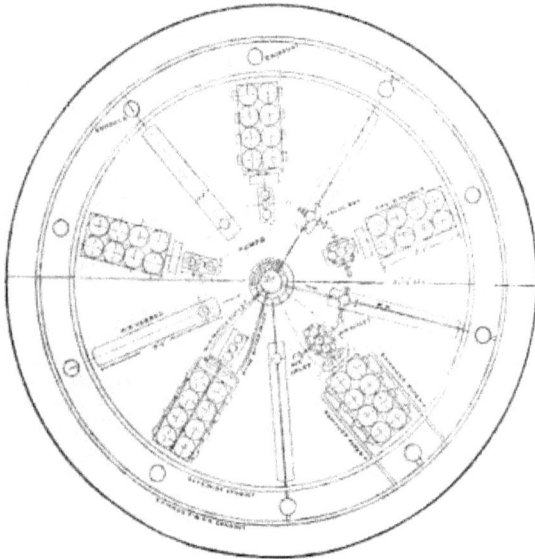

PLAN OF PITHA (BASE)

Drawn by
T. K. ELLAPPA,
Bangalore.
1-12-1923.

Prepared under instruction of
Pandit SUBHARAYA SASTRY,
of Anekal, Bangalore

SUNDARA VIMANA

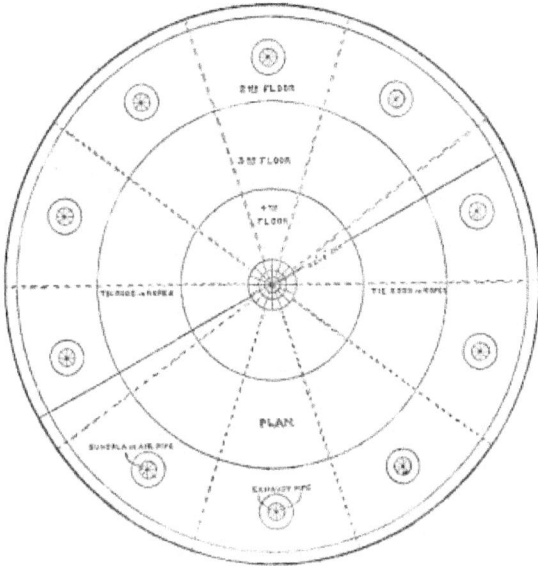

FLOORS

Drawn by
T. K. ELLAPPA,
Bangalore.
2-12-1923.

Prepared under instruction of
Pandit SUBBARAYA SASTRY,
of Anekal, Bangalore.

SUNDARA VIMANA

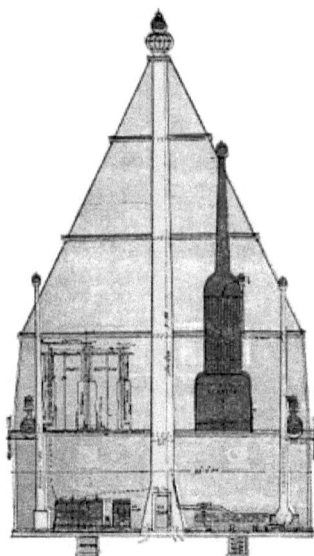

VERTICAL SECTION

Drawn by
T. K. ELLAPPA,
Bangalore.
2-12-1923.

Prepared under instruction of
Pandit SUBBARAYA SASTRY,
of Anekal, Bangalore

TRIPURA VIMANA

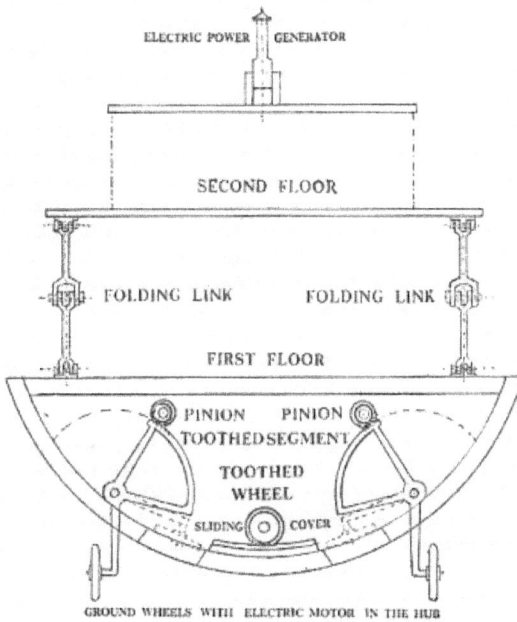

ELECTRIC POWER GENERATOR

SECOND FLOOR

FOLDING LINK   FOLDING LINK

FIRST FLOOR

PINION   PINION
TOOTHED SEGMENT

TOOTHED
WHEEL

SLIDING   COVER

GROUND WHEELS WITH ELECTRIC MOTOR IN THE HUB

CROSS SECTION

Drawn by
T. K. ELLAPPA,
Bangalore.
3-12-1923.

Prepared under instruction of
Pandit SUBBARAYA SASTRY,
of Anekal, Bangalore

ELECTRIC POWER GENERATOR

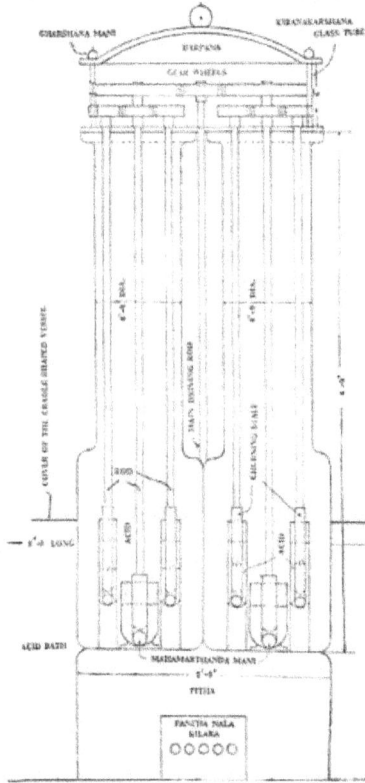

SECTIONAL ELEVATION

Drawn by
T. K. ELLAPPA,
Bangalore.
2-12-1923.

Prepared under instruction of
Pandit SUBBARAYA SASTRY,
of Anekal, Bangalore

ELECTRIC POWER GENERATOR
TOP VIEW

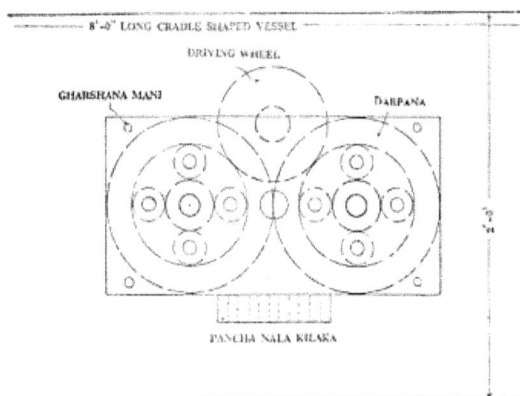

——————— 8'-0" LONG CRADLE SHAPED VESSEL ———

DRIVING WHEEL

GHARSHANA MANI   DARPANA

PANCHA NALA KILAKA

ELECTRIC   MOTOR

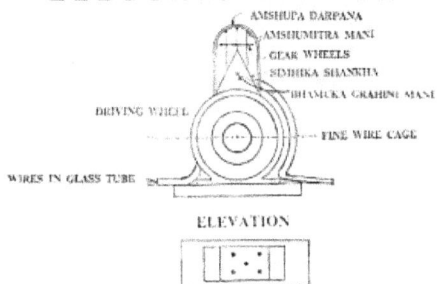

AMSHUPA DARPANA
AMSHUMITRA MANI
GEAR WHEELS
SIMHIKA SHANKHA
BHAMUKA GRAHINI MANI

DRIVING WHEEL

FINE WIRE CAGE

WIRES IN GLASS TUBE

ELEVATION

PLAN

Drawn by
T. K. ELLAPPA,
Bangalore.
1-12-1923.

Prepared under instruction of
Pandit SUBBARAYA SASTRY,
of Anekal, Bangalore

TRIPURA VIMANA

PERSPECTIVE VIEW

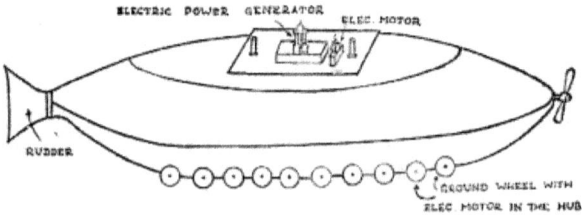

ELECTRIC POWER GENERATOR   ELEC. MOTOR

RUDDER

GROUND WHEEL WITH
ELEC MOTOR IN THE HUB

VERTICAL SECTION

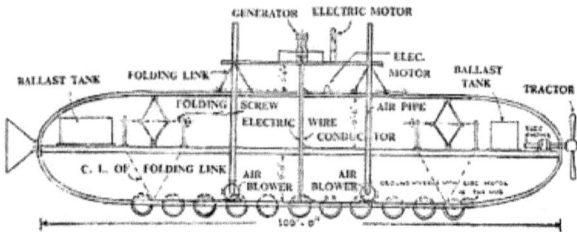

GENERATOR ELECTRIC MOTOR

ELEC.
MOTOR

BALLAST TANK   FOLDING LINK   BALLAST
                                          TANK   TRACTOR

FOLDING   SCREW
FOLDING   ELECTRIC WIRE   AIR PIPE
               CONDUCTOR

C. L. OF FOLDING LINK   AIR   AIR
                          BLOWER   BLOWER

100'- 0"

PLAN

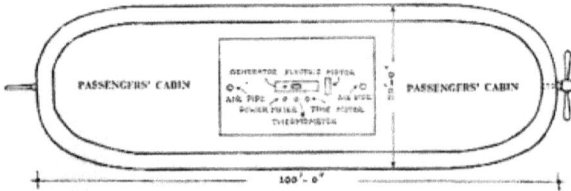

PASSENGERS' CABIN   GENERATOR ELECTRIC MOTOR   PASSENGERS' CABIN

AIR PIPE   AIR PIPE
POWER METER   TIME METER
THERMOMETER

100'- 0"

SHAKUNA VIMANA

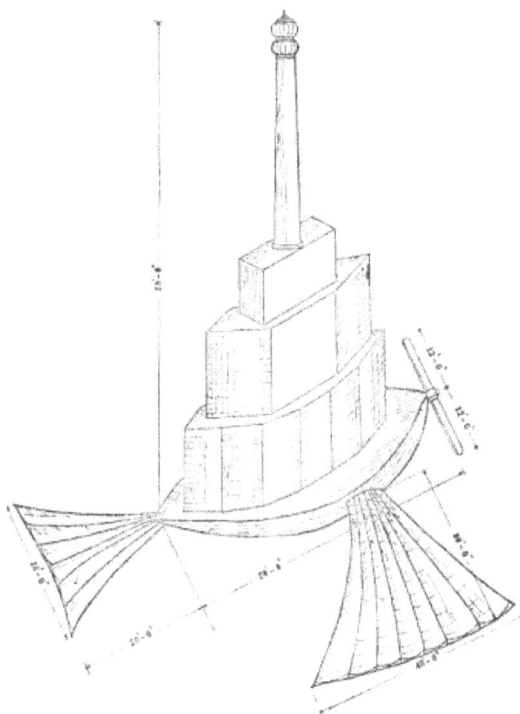

PERSPECTIVE VIEW

Drawn by
T. K. ELLAPPA,
Bangalore.
2-12-1923.

Prepared under instruction of
Pandit SUBBARAYA SASTRY,
of Anekal, Bangalore.

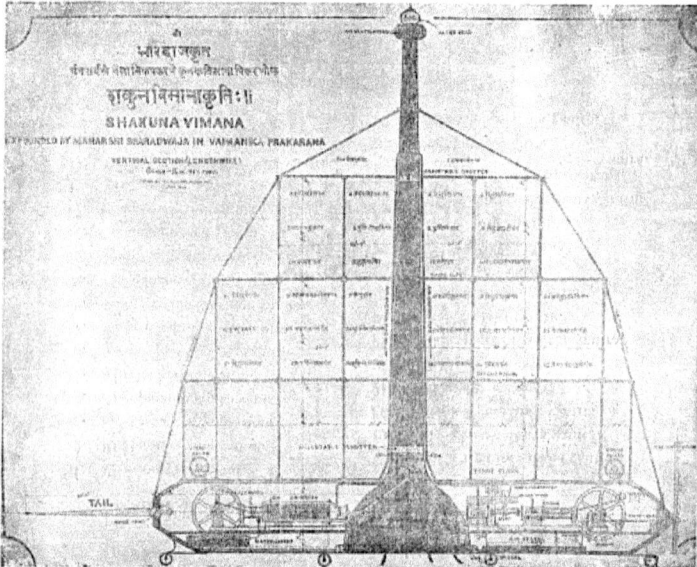

SHAKUNA VIMANA

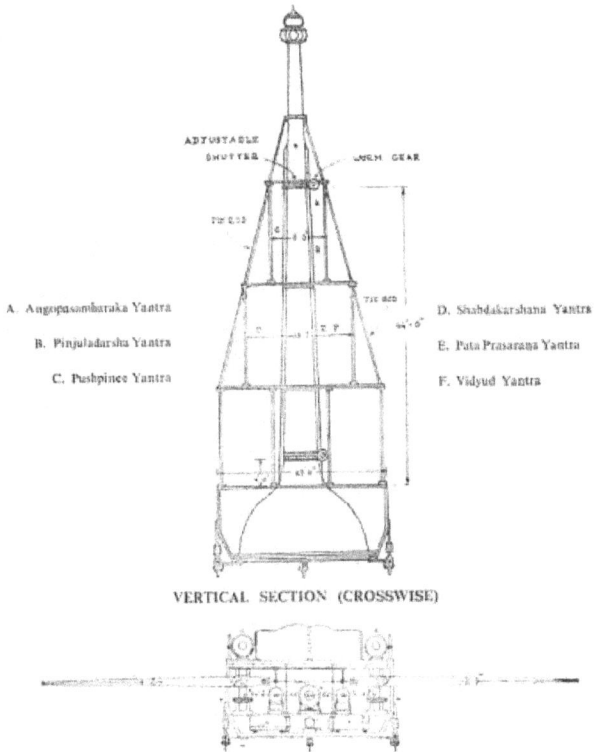

ADJUSTABLE
SHUTTER   WORM GEAR

A. Angopasamharaka Yantra

B. Pinjuladarsha Yantra

C. Pushpinee Yantra

D. Shabdakarshana Yantra

E. Pata Prasarana Yantra

F. Vidyud Yantra

VERTICAL SECTION (CROSSWISE)

LEGEND :   G. Steam Boiler H. Engine J. Oil Tank K. Air Vessel
L. Air Heater P. Pivot T. Turning Wheel W. Wing

VERTICAL SECTION AT THE WING JOINT

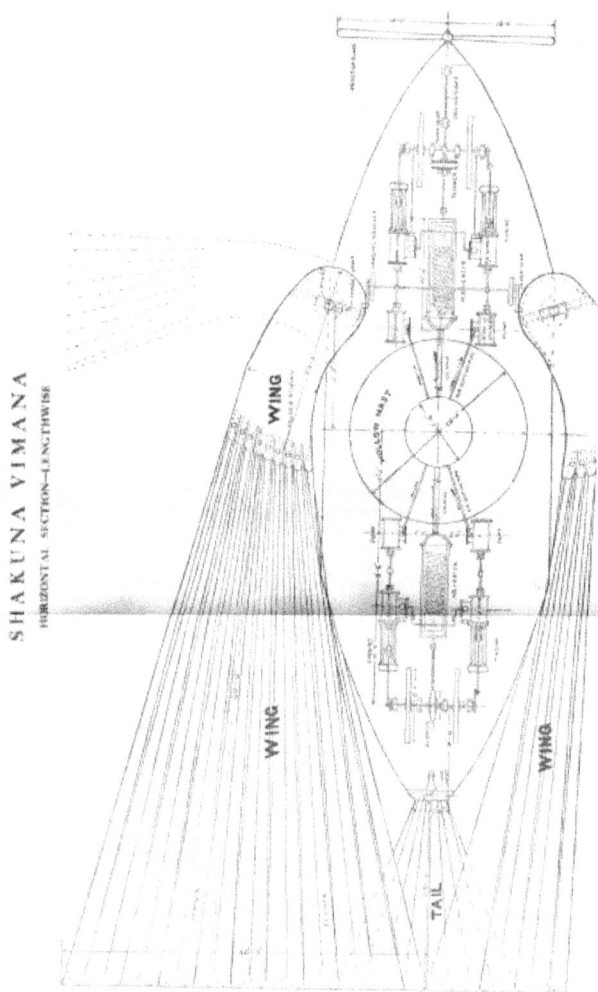

SHAKUNA VIMANA
HORIZONTAL SECTION—LENGTHWISE

WING

WING

WING

TAIL

Edizioni L'Arcipelago
Via Panciatichi, 23/6
50100 Firenze
tel: 3408135490

Stampato per conto di
Edizioni L'Arcipelago
prima stampa: agosto 2017